T0202218

Objective Becoming

What does the passage of time consist in? There are some suggestive metaphors. "Events approach us, pass us, and recede from us, like sticks and leaves floating on the river of time." "We are moving from the past into the future, like ships sailing into an unknown ocean." There is surely something right and deep about these metaphors. But how close are they to the literal truth? In this book Bradford Skow argues that they are far from the literal truth. Skow's argument takes the form of a defense of the block universe theory of time, a theory that, in many ways, treats time as a dimension of reality that closely resembles the three dimensions of space.

Opposed to the block universe theory of time are theories that take the metaphors more seriously: presentism, the moving spotlight theory, the growing block theory, and the branching time theory. These are theories of "robust" passage of time, or "objective becoming." Skow argues that the best of these theories, the block universe theory's most worthy opponent, is the moving spotlight theory, the theory that says that "presentness" moves along the series of times from the past into the future. Skow defends the moving spotlight theory against the objection that it is inconsistent, and the objection that it cannot answer the question of how fast time passes. He also defends it against the objection that it is incompatible with Einstein's theory of relativity. Skow proposes several ways in which the moving spotlight theory may be made compatible with the theory of relativity.

Still, this book is ultimately a defense of the block universe theory, not of the moving spotlight theory. Skow holds that the best arguments against the block universe theory, and for the moving spotlight theory, start from the idea that, somehow, the passage of time is given to us in experience. Skow discusses several different arguments that start from this idea, and argues that they all fail.

Objective Becoming

Bradford Skow

OXFORD
UNIVERSITY PRESS

OXFORD
UNIVERSITY PRESS

Great Clarendon Street, Oxford, OX2 6DP,
United Kingdom

Oxford University Press is a department of the University of Oxford.
It furthers the University's objective of excellence in research, scholarship,
and education by publishing worldwide. Oxford is a registered trade mark of
Oxford University Press in the UK and in certain other countries

First published 2015
First published in paperback 2016

Published in the United States of America by Oxford University Press
198 Madison Avenue, New York, NY 10016, United States of America

British Library Cataloguing in Publication Data
Data available

Library of Congress Cataloging in Publication Data
Data available

ISBN 978-0-19-871327-2 (Hbk.)
ISBN 978-0-19-877669-7 (Pbk.)

Acknowledgments

When I write my second book I expect I will want to thank all the people who helped me write it. But this is my first book, and I am inclined to thank *everybody*, including the people who helped put me in a position to write philosophy books in the first place. I will try to keep that impulse in check, but some people cannot go unmentioned.

I owe a great deal to Eldra Avery for teaching me how to write, to Dan Merrill for telling me I am good at philosophy, to Kathie Linehan for support in a difficult time, and to Bob Longsworth for many years of encouragement, advice, and love, even after I abandoned his chosen field of study. I probably owe the most to Fred Feldman, who has taught me more than anyone about how to do philosophy, and how to be a philosopher.

Fred, as it turns out, read parts of this book and sat through a seminar presentation on it, so he is a good bridge to the people who helped specifically with this project. Oliver Pooley wrote detailed, and extremely helpful, comments on an early draft. Yuri Balashov and his students suffered through that same draft in his spring, 2013 seminar, and I am grateful for their feedback. Maya Eddon and the students in her fall, 2013 seminar benefited from an improved draft, and also helped improve it themselves. Gordon Belot and the students in his winter, 2014 seminar read the penultimate draft and did everything from find typographical errors to raise difficult questions that I am not sure I have adequately addressed. Among the longest sets of comments and criticisms I received were from three referees (Simon Prosser was one, the others remain anonymous). I am grateful for their efforts; the book is better for them.

I must thank Ted Sider for generously taking the time to read and respond to my work, including this manuscript, over many years. Several times I had what I thought was a brilliant and brilliantly original insight, only to then find it somewhere in Ted's writing.

Many other people read large chunks of the manuscript, and told me what they thought: a million thanks to Phil Bricker, Chris Meacham, Jack Spencer, Melissa Schumacher, Storrs McCall, Brad Weslake, Chris Heathwood, Shamik Dasgupta, Ned Markosian, Kris McDaniel, Steve Yablo, Josh Schechter, Michael Huynh, and Damiano Costa. I presented some of this material in a summer seminar at MIT in 2013, and owe a lot to those who came: Agustin Rayo, Caspar Hare, Bernhard Salow, Sofia Ortiz-Hinojosa, Dylan Bianchi, Matthias Jenny,

Matt Mandelkern, Abbey Jacques, Brendan Dill, Dave Gray, and Rose Lenehan (I hope I didn't miss anyone!). Thanks also to the audience at USC. A lot of the work on this book was supported by a Charles A. Ryskamp Research Fellowship from the American Council of Learned Societies.

Finally, my family. To my parents for never flinching when I said I would become a philosopher, to my birth mother Mary for eighteen years of being there (how time flies), to my sons Elliot and Nathaniel for all the smiles and laughs at the end of the day (not to mention the countless renditions of "Yellow Submarine" with toy ukelele accompaniment), and to my wife Deanna for taking care of us all.

Traces of several of my papers appear in several parts of this book, though for the most part everything has been rewritten. In particular, Chapters 11 and 12 are a significant expansion and revision of "Experience and the Passage of Time." I have changed my mind about several things since writing that paper. I thank the publishers for letting me use material from the following papers:

With kind permission from Springer Science+Business Media: "On the Meaning of the Question 'How Fast Does Time Pass?'," *Philosophical Studies* 155, 2011, 325–44.

With permission from *The Journal of Philosophy*: "Relativity and the Moving Spotlight," 106, 2009, 666–78.

With permission from John Wiley & Sons: " 'One Second Per Second'," *Philosophy and Phenomenological Research* 85, 2012, 377–89; and "Experience and the Passage of Time," *Philosophical Perspectives 25: Metaphysics*, 2011, 359–87.

Contents

List of Figures   ix

1. Introduction: Time Passes?   1

2. The Block Universe   4
   2.1 Spacetime   4
   2.2 The Picture and the Theory   10
   2.3 The Denial of Passage   17
   2.4 Appendix: What the Block Universe Theory is Not   18
   2.5 Appendix: Other Versions of Reductionism about Tense   19

3. What Might Robust Passage Be?   22
   3.1 Passage and "Real Change"   22
   3.2 Presentism   27
   3.3 Passage and Presentism   32
   3.4 Two Metaphors for Passage   36
   3.5 Appendix: More on Passage and Presentism   38

4. The Moving Spotlight   44
   4.1 Can Time Itself Move or Flow?   44
   4.2 The Moving Spotlight via Supertense   49
   4.3 The Moving Spotlight via Temporal Perspectives   58
   4.4 Appendix: MST-Time with Absolute Presentness   68
   4.5 Appendix: MST-Time and Semantic Relativism   68

5. Growing Blocks and Branching Times   70
   5.1 Two More Theories of Robust Passage   70
   5.2 The Open Future   72
   5.3 Appendix: A Remark on the Structure of Theories of Time   81

6. The Moving Spotlight Theory is Consistent   82
   6.1 The Nuclear Option   82
   6.2 McTaggart   83
   6.3 McTaggart Redux: Smith   88
   6.4 McTaggart Redux: Mellor   90
   6.5 McTaggart Redux: Van Cleve   95

7. How Fast Does Time Pass?   101
   7.1 An Unanswerable Question?   101
   7.2 Changes without Rates of Change   103
   7.3 Seconds and Superseconds   109
   7.4 Arguments from Dimensional Analysis   114
   7.5 Odds and Ends   127

List of Figures

2.1.   Space and time on the 3 + 1 view.   5

2.2.   Spacetime on the 4D view.   6

2.3.   Motion on the 3 + 1 view.   6

2.4.   Instantaneous copies of space at each time.   7

2.5.   Motion through instantaneous spaces.   7

2.6.   Space and time in the 4D view.   8

2.7.   A picture of spacetime and its contents.   10

2.8.   The motion of a clock hand in the block universe.   17

3.1.   The motion of a clock hand in the block universe.   23

3.2.   A picture of spacetime and its contents.   27

3.3.   Space and time on the 3 + 1 view.   29

3.4.   Spatiotemporal reality according to presentism.   29

3.5.   A presentist diagram.   31

3.6.   More presentist diagrams.   31

3.7.   The passage of time in the block universe.   35

3.8.   The passage of time in presentism.   35

4.1.   Time in the block universe.   45

4.2.   The present in the moving spotlight theory.   45

4.3.   The moving spotlight with supertime.   46

4.4.   Time moving.   48

4.5.   Schlesinger's moving spotlight theory.   52

4.6.   Temporal reality in MST-Supertense.   53

4.7.   Adventures of a square.   55

4.8.   Further adventures of a square.   56

4.9.   A problem for the strong truth-conditions?   58

4.10. How reality is, from various perspectives in supertime.   60

4.11. Pictures of temporal reality in MST-Time.   61

4.12. Two viewers of a cylinder.   63

4.13.	The cylinder from H's perspective.	63
4.14.	The cylinder from V's perspective.	63
5.1.	The growing block.	71
5.2.	A picture of branching time.	72
5.3.	Passage in branching time.	72
5.4.	The future in the three theories.	73
5.5.	The open future in the moving spotlight theory.	75
5.6.	A branching time scenario.	79
6.1.	Pictures of temporal reality in MST-Time (again).	89
6.2.	Two sentence tokens in the block universe.	91
6.3.	Two sentence tokens in MST-Time.	93
6.4.	Two sentence tokens in MST-Supertense.	95
7.1.	Incorrect passage.	108
7.2.	Correct passage.	108
8.1.	Space and time in Newtonian spacetime.	131
8.2.	The structure of Minkowski spacetime.	132
8.3.	A worldline in Newtonian spacetime.	133
8.4.	Worldlines in Minkowski spacetime.	134
8.5.	Proper time in Minkowski spacetime.	136
8.6.	Relative temporal separation in Minkowski spacetime.	137
8.7.	Relativistic right angles.	138
8.8.	Relative simultaneity in Minkowski spacetime.	139
8.9.	The spotlight in Newtonian spacetime.	142
9.1.	One spotlight in Minkowski spacetime.	147
9.2.	A spacelike line.	151
9.3.	Present times from two points in supertime.	152
9.4.	Points that are never present.	154
9.5.	Two neo-classical theories.	157
9.6.	A deviant classical theory.	158
9.7.	A solipsistic neo-classical theory.	159
9.8.	A superspacetime theory.	162
9.9.	Past, present, and future from the perspective of time T.	165

9.10.   Past, present, and future from the perspective of P.   166

9.11.   Q becomes past without ever being present?   168

10.1.   Me, moving through time.   179

10.2.   Me, in the block universe.   180

11.1.   The Müller-Lyer illusion.   190

11.2.   The brown star looks red.   196

11.3.   The star looks present?   197

11.4.   It super-was the case that . . .   198

11.5.   The red rooms according to MST-Supertense.   204

12.1.   My meditations according to the block universe.   208

12.2.   My meditations according to MST-Supertense.   208

12.3.   MST-Supertense with time-independent availability.   215

12.4.   The stage theory.   217

12.5.   My meditations according to the block universe.   220

12.6.   My meditations according to the block universe theory +
        the stage theory.   220

12.7.   My meditations according to the spatial block universe.   222

12.8.   My meditations according to the spatial moving spotlight theory.   222

12.9.   Some worlds in Jane's relief state according to MST-Supertense.   229

1

Introduction: Time Passes?

As Dylan Thomas's *Under Milk Wood* begins, all the townspeople are asleep but we are awake. We hear the dew falling and the "invisible starfall" and the "darkest-before-dawn minutely dewgrazed stir of the black, dab-filled sea." Then the First Voice, who narrates the opening, says: "Time passes. Listen. Time passes." It is a moment of great anticipation. Immersed in the world of the play it is natural to ask, What is coming?

But a philosopher who holds himself outside of the world of the play may ask a different question. He may ask, Just what is this process—the passage of time—that (supposedly) goes on as we listen?

We often reach for metaphors when we describe the passage of time. We say that events are carried along by the passage of time from the future into the past, like sticks and leaves floating on a river. Or we say that we move through time from the past into the future, like ships sailing on an unknown sea. But how seriously should we take these metaphors?

In my view, not very seriously. If the passage of time requires that time move or flow in anything like the way rivers do then, I think, there is no such thing as the passage of time. If the passage of time requires that we move through time in anything like the way that trains move through space, then there is no such thing as the passage of time. I hold that time is a lot like space, "just another dimension." Of course I think there are differences between the time dimension and the three dimensions of space.[1] I just deny that there is some mysterious process, "passage," that time undergoes but space does not.

The claim that time is a lot like space is vague, because it says nothing about which aspects of time and space are similar, or about how similar they are. The theory of time in which this claim of similarity is made more precise is sometimes

---

[1] In "What Makes Time Different From Space?" I defended the claim that it is the role that time plays in the laws of nature that makes it different. I am not sure I still believe this view, but it is compatible with denying that time passes.

called the "block universe" theory of time.[2] This book is a defense of the block universe theory's account of the passage of time.

In some contexts it is natural to say that, according to the block universe theory, time does not pass. But in other contexts this is a misleading thing to say. For one thing, it can make the block universe theory sound like a mystical doctrine. Some say "time does not pass" to mean that the temporal aspects of reality are a complete illusion. But the block universe theory is not a form of mysticism. Nor do those of us who accept it think that ordinary people are constantly saying false things when they say that time is flying by, or that the passage of time has got them feeling down. The people who we think say false things are the philosophers who defend alternative theories of the passage of time, not non-philosophers who have given no thought to which theory is correct.

To avoid misleading people it is better to allow that in a sense time passes if the block universe theory is true, but that that passage is "anemic." The theory lacks "robust" passage of time. This raises two questions. First, if the block universe theory lacks robust passage then what might robust passage be? What theories of time are there that do contain robust passage?

Once we know what the alternatives are we can go on to ask the second question. What kind of theory of time should we believe? One with robust passage, or one without? My answer to the second question is, of course, that we should believe the block universe theory. But my argument for this answer will follow an unusual path.

The first thing I am going to do is identify my most worthy opponents, the best theories of robust passage. My sense is that many philosophers think that the block universe theory's most natural opponent is a theory of time called presentism. But I will argue in Chapter 3 that if it is robust passage you are after then presentism is not the right theory for you. The best theory of robust passage, I will argue, is the moving spotlight theory of time.

While I believe the block universe theory, my relationship with the moving spotlight theory is, as they say, complicated. I think the theory is fantastic. That is, I think it is a fantasy. But I also have a tremendous amount of sympathy for it. Not in the sense that I want it to be true, but in the sense that I find it easy to think from the position of someone who accepts it. (Or so I believe; I have not

---

[2] Sometimes the block universe theory is called "the B-theory of time." But it is just a coincidence that "block" starts with a "b." The term "B-theory" comes from a distinction McTaggart made, which I will discuss in Chapter 6. One of the earliest defenders of the block universe theory was Bertrand Russell; see for example his 1915 paper "On the Experience of Time." There have been many others since. A recent defense, against a somewhat different opponent than mine, with extensive references, is in chapter 2 of Ted Sider's book *Four-Dimensionalism*.

had this statement verified by someone who does accept it. Maybe they will find this statement incredible.) Now philosophers on both sides of the aisle—those who believe in robust passage and those who do not—have held that the moving spotlight theory is easily refuted. Many think that contemplating the moving spotlight theory is like watching a bad time-travel movie. Each short stretch of the movie might make sense on its own, but if you think about the movie as a whole there are contradictions everywhere. It is also common to think that the moving spotlight theory runs into trouble as soon as you ask how fast time passes, or that it relies on a way of thinking about time that Einstein's theory of relativity shows to be false. I do not think any of these strategies for refuting the moving spotlight theory succeed. And I will spend Chapters 6 through 9 defending the moving spotlight theory against them. This may seem strange, since my ultimate aim is to argue that the theory should be rejected. But philosophy is not politics. I don't care why you vote for my candidate, but I want you to believe the block universe theory for the right reasons.

I think that the endgame in the debate between the block universe theory and the moving spotlight theory is over how the theories make sense of our experience of time. I think that the strongest case in favor of the moving spotlight theory starts with the claim that it explains some features of our experience better than the block universe does. I will examine that case in Chapters 11 and 12.

2

The Block Universe

2.1 Spacetime

"Spatiotemporal reality is nothing but a four-dimensional block universe." So says the block universe theory, but it is not very clear what this statement means. I want to work toward a clearer statement. To do that we are going to need to know what spacetime is. (If you already know what spacetime is, please bear with me.)

A naive view about space and time goes like this. Space is a three-dimensional thing made up of points. Points of space are the smallest parts of space, and they have zero size. Similarly, time is a one-dimensional thing made up of instants. Instants of time are the smallest parts of time, and they also have zero size (zero temporal extent). So on this view the basic spatiotemporal things are points of space and instants of time. Points of space and instants of time are different kinds of thing, and space and time are "separately existing" things. I will call this the 3 + 1 view. In a picture the 3 + 1 view looks like Figure 2.1 (which, for convenience, depicts space as two-dimensional).

Plenty of aspects of the 3 + 1 view are controversial. It is a version of substantivalism about space and time. It says that space and time exist. An alternative view, relationalism about space and time, says they do not. But my presentation of the block universe theory will presuppose substantivalism.[1] (Relationalists do not mean to say that it is a complete illusion that, for example, I am four feet from that table. While they deny that there are any such things as the regions of space that I, and the table, occupy, they also reject the idea that the table and I can be four feet apart only by occupying regions of space that are, in a more direct way, four feet apart.)

The 3 + 1 view as I stated it also makes claims about the structure of space and time that have been denied. Some philosophers say that space exists, and

---

[1] Later I will describe a theory called presentism which says that time does not exist. But presentism and relationalism are very different theories. There are versions of the block universe theory that assume relationalism about space and time, but presentism is incompatible with the block universe theory.

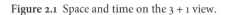

Figure 2.1 Space and time on the 3 + 1 view.

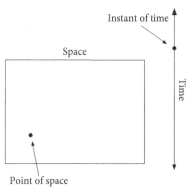

has parts, but deny that there are smallest parts of space and so deny that there are points of space. They say something similar about time. I will also assume that this view is false. I am making these assumptions—substantivalism and "pointism"—without defending them not because I think that they need no defense, but because I find them natural assumptions to work with, and defending them would take me too far from the main line of argument.

So that is the 3+1 view. I have called it a "naive" view because it is a comfortable, natural, and easy-to-grasp way of thinking about space and time that physics has largely abandoned. The 3 + 1 view is compatible with the official statement of the block universe theory, and a lot of my discussion of the passage of time in the first several chapters of this book will assume it. But the picture that goes with the block universe theory presupposes a different view about space and time, namely the 4D view.

The mathematician Hermann Minkowski was thinking of the 4D view when he famously wrote that

space by itself, and time by itself, are doomed to fade away into mere shadows, and only a kind of union of the two will preserve an independent reality. ("Space and Time," p. 75.)

As this quotation suggests, the 4D view rejects the claim that points of space and instants of time are the basic spatiotemporal things. It says instead that the basic spatiotemporal things are points of spacetime, and that spacetime itself, the thing made up of these points, is a four-dimensional thing. Figure 2.2 contains a three-dimensional representation of four-dimensional spacetime.

Fine; but what is spacetime? If you do not already know this picture is not going to help.

Go back to the 3 + 1 view. What is space on the view? Space is the "arena" in which spatiotemporal things (material bodies, like hawks and handsaws) exist

Point of spacetime

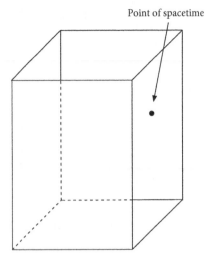

Figure 2.2 Spacetime on the 4D view.

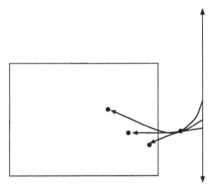

Figure 2.3 Motion on the 3 + 1 view.

and move around. Now a body that moves around does not just have one single spatial location. It is in different places at different times. A diagram of how something has moved around must indicate where it is in space at each time. Figure 2.3 is a diagram like that that uses the 3 + 1 view. The dot floating between time and space in the diagram represents a very small material body. An arrow from an instant of time that passes through the dot points to the region of space that body occupies at that time. (The diagram includes only three arrows, though of course really the body exists during a continuous stretch of time that includes those three.)

Now in this picture—and in the 3 + 1 view—points of space always exist. But we could think of things differently. Instead of having a single thing, space, that always exists, suppose instead that there are infinitely many copies of space, each

Figure 2.4 Instantaneous copies of space at each time.

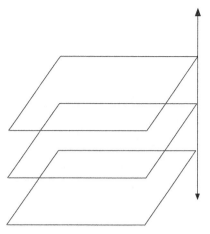

Figure 2.5 Motion through instantaneous spaces.

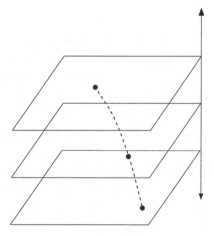

of which exists for just one instant. Place each copy of space next to the instant at which it exists. The picture that results might look like Figure 2.4.

Now to describe something's motion we just need to indicate where it is in each "instantaneous space" at the time connected to that instantaneous space. That is easy to do by just placing dots on each instantaneous space, as in Figure 2.5.

But now notice that there is no need to believe that in addition to the stack of instantaneous spaces there also exists instants of time. We could instead just identify instants of time with those instantaneous spaces. Then we have arrived at the 4D view. The stack of instantaneous spaces just is spacetime. The smallest parts of the stack are "instantaneous points of space." But an instantaneous point of space is something with zero spatial extent and zero temporal extent. And that

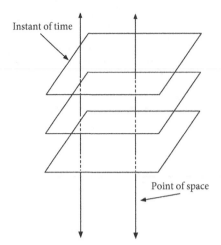

Instant of time

Point of space

Figure 2.6   Space and time in the 4D view.

is what points of spacetime are supposed to be like. So the points of instantaneous space are points of spacetime, which are the smallest parts of spacetime.

We have seen what times are on the 4D view. What about points of space? Just as instants of time are "instantaneous spaces," which are three-dimensional slices of spacetime, so points of space are "local times," one-dimensional lines through spacetime. The idea is easiest to convey with a diagram, so look at Figure 2.6.

Now in my "construction" of spacetime I stacked up infinitely many instantaneous spaces. I said that each instantaneous space was a copy of space. What exactly does that mean? It certainly means that each instantaneous space has the same geometry that space has in the 3 + 1 view. But I should point out that Figure 2.6 makes sense only if there is more to be said about the relationship between the instantaneous spaces than that they have the same geometry. If all we can say is that each instantaneous space has the same geometry, say Euclidean geometry, then while we can speak of the spatial distance between two points in one single instantaneous space, there would be no sense to be made of talk of the spatial distance between points from different instantaneous spaces. Figure 2.6 requires this later kind of talk to make sense. So assume that for any pair of spacetime points X and Y there is a number that is the spatial distance (in, say, meters) between X and Y. (And assume that these numbers obey some obvious laws. Assume that, given a point X and an instantaneous space S, there is exactly one point Y on S that is zero meters from X. And assume that if X is zero meters from Y, and Y is N meters from Z, then X is also N meters from Z.) Then we can identify points of space with certain lines in spacetime. Pick a spacetime point on one of the instantaneous spaces. Then look at another nearby instantaneous space, and find the point on it that is zero spatial distance from the point you started with. The line in spacetime connecting these two points, and passing through all and only the other points on other instantaneous spaces that are zero spatial distance from the point you started with, is a point of space. These lines behave just like points of space do on the 3 + 1 view. As in that view, each point of space exists at each time, since each line that is identified with a point of space passes through each instantaneous space. As in that view, something that is, at each time, located at the same point of space does not move.

The relationship between the 4D view, as I have described it so far, and the 3 + 1 view is a little subtle. Both views say that points of space and instants of time exist. What the views differ on is their nature. The 3 + 1 view says that neither points of space nor instants of time have any parts, and that they are different kinds of thing. But the 4D view says that they do have parts. Each is made of up points of spacetime. So there is still such a thing as space; but there is no such thing as space "by itself," something completely separate from time. For now places and times are made of the same things, points of spacetime, and any point of space and instant of time have a spacetime point in common.

The way I have presented the 4D view so far is misleading in one respect. We are used to the idea that there are many different geometries that space may have. There is Euclidean geometry, but also spherical geometry and hyperbolic geometry. It is also true that there are many different geometries that spacetime may have. I have presented the 4D view with one particular spacetime geometry in mind, namely Newtonian spacetime. But there is excellent evidence that spacetime does not in fact have this geometry. And in most of the other geometries spacetime might have, including the ones that contemporary physics takes seriously, it is wrong to think of spacetime both as a stack of instantaneous spaces and a bundle of local times.[2] That is, in some of those spacetime geometries points of space do not exist, and in some of them neither points of space nor instants of time exist. In those geometries there is no such thing as "being in the same place at different times." But even if it has one of these other geometries spacetime still plays the same role. To describe something's spatiotemporal career is to say which region of spacetime it occupies. And space and time are not completely gone in these geometries. There are still facts about what "counts as" a point of space or an instant of time relative to an observer in spacetime.[3] These other

---

[2] Thus when doing physics the 3+1 view is not tenable. For arguments that the 4D view is superior to the 3 + 1 in philosophy, not just in physics, see John Earman's paper "Space-Time, or How to Solve Philosophical Problems and Dissolve Philosophical Muddles Without Really Trying."

[3] In some relativistic spacetimes there is no way to divide up all of spacetime into the times-relative-to-observer-X (similarly for points of space), but there is a way to do this locally.

geometries will be important later, when we talk about the bearing of the theory of relativity on debates about the passage of time, but I will not say anything more about them now.

2.2 The Picture and the Theory

I said earlier that according to the block universe theory spatiotemporal reality is nothing but a four-dimensional block universe. There is a picture that goes with this theory, and also a somewhat more detailed statement of the theory in words. While the statement is more detailed, I think the picture guides people's thinking about the theory as much as, maybe more than, the statement. So I will start with the picture.

The picture is just a picture of spacetime and its contents. If we imagine that there are only three very small material things in the whole universe, then a picture of spacetime and its contents might look like the picture in Figure 2.7. (For convenience I have drawn spacetime as two-dimensional and labeled two times, T1 and T2. Material bodies A and B are close together at T1, then by T2 have moved farther apart.) The idea that goes with the picture is that a picture like this can be a complete picture of spatiotemporal reality.

A statement of the block universe theory starts with a claim that is evident from the picture. There are such things as times (or, at least, there is such a thing as spacetime). This claim is often called eternalism. The theory goes on to endorse a thesis called reductionism about tense, which I shall now explain.

As a first pass, reductionism about tense is the claim that a complete description of reality can be given from an "atemporal perspective." Now according to

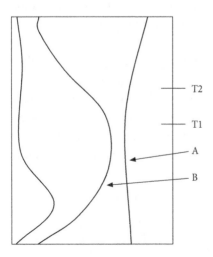

Figure 2.7 A picture of spacetime and its contents.

T2

T1

A

B

the block universe theory, sentences with tensed verbs do not make sense from an atemporal perspective. A sentence like "George climbed the tree" is true at some times, false at others; it lacks a "time-independent" truth-value. (Again, this is something that the block universe theory says about tensed language. We will later encounter a theory of time that denies it.)

However, the claim that

(1) If reductionism about tense is true, then a complete description of reality can be given without using any tensed verbs.

is only a partial characterization of reductionism about tense. The converse of (1) is not true. Reductionism about tense is not equivalent to the claim that a tenseless description can be complete.

Before explaining why let me say something about tense. English contains sentences that have tensed verbs. "George climbed the tree," for example, is in the past tense, and "George is hungry" is in the present.[4] But I do not myself know whether English contains any sentences that have only tenseless verbs. One might think that "is" in "George is hungry at noon on May 2, 2012" is tenseless. But one might also doubt that this is really a well-formed sentence of English.[5] As far as reductionism about tense is concerned it does not matter whether English has tenseless verb forms. If it does not then the complete descriptions of reality that reductionism speaks of cannot be descriptions in English. They must belong to some other language, perhaps some language that no one speaks. To keep things simple I will assume that sentences like "George is hungry at noon on May 2, 2012" are tenseless. (If these sentences are not English then we may assume that they belong to another language that expands English by adding tenseless verb forms.) It should usually be clear from the context whether a verb

[4] Tense is a very complicated part of English grammar. I know relatively little about it. In *A Student's Introduction to English Grammar* Huddleston and Pullum emphasize that "the relation between tense and time in English is not at all straightforward" (p. 44). The correct name for what I have called the "past tense" form of a verb is the "preterite." The preterite is often used to talk about the past, but not always ("I'd rather she arrived tomorrow" is one of Huddleston and Pullum's examples (p. 46)). Also, technically there is no such thing as the future tense form of a verb in English (p. 56). Instead "the most basic way" to get an English sentence about the future is to add the modal auxiliary "will" to a bare infinitival clause containing the "plain form" of a verb. ("George will feed the rabbit" is like this.) This shows that tense is not the only grammatical category relevant to how English "locates things in time." Aspect and mood are also important. I will for the most part ignore these complications.

[5] English does contain tenseless verb forms. The plain form of a verb is an example. The occurrence of "sing" in "I heard George sing" is the plain form. But what the reductionist needs are tenseless sentences, not merely tenseless verb forms. The plain form of a verb cannot be the only verb in a sentence. In the example just given there is also the verb "heard," which is tensed. For an argument that English does contain tenseless sentences see Zimmerman, "The A-Theory of Time, the B-Theory of time, and 'Taking Tense Seriously'."

in a sentence I write is meant to be tenseless or not. When I want to make the distinction between present-tense and tenseless verbs explicit I will underline verbs that should be read as tenseless.[6] So the "is" in "George <u>is</u> hungry at noon on May 2, 2012" is tenseless.

Returning to the main thread, why is reductionism about tense not equivalent to the claim that a tenseless description can be complete? I said that the idea behind reductionism is that a complete description of reality can be given from an atemporal perspective. But according to the block universe theory, there are sentences that are tenseless but still do not make sense from an atemporal perspective. "George <u>finishes</u> climbing the tree at a moment of past time" is an example. The block universe theory says that this sentence may be true at some times, false at others, but that it lacks a time-independent truth-value. So a more precise statement of reductionism about tense looks something like this:

(2) A complete description of reality can be given without using tense and without using the words "present," "past," "future," "now," "tomorrow," and so on.

A complete list of the banned words and phrases would also have to include, for example, "last Thursday" and "the next time I see you"; it would also have to ban pieces of vocabulary in other languages that are synonymous with these. The complete list may well be infinite. To have a name for the items of vocabulary on this list I will call it the A-vocabulary.[7] (I call it this for the sake of tradition and because I can think of no better name.)

It would be nice to have a way to specify the A-vocabulary other than by writing down a long list. One natural idea is that the A-vocabulary is vocabulary whose presence in a sentence prevents that sentence from having a time-independent truth-value. But this is not a good way to define "A-vocabulary." For we want all theories of time to be able to agree on what the A-vocabulary is. But not all theories of time agree that, for example, the presence of "past" in a sentence can prevent that sentence from having a time-independent truth-value. (Presentism, which I will discuss in Chapter 3, is a theory that disagrees with this.)

The following might sound like a better characterization of the A-vocabulary. The A-vocabulary is the vocabulary the presence of which prevents sentences

---

[6] This typographical convention is inspired by Szabó, who adopts a similar one in "Counting Across Times." Others, including Falk and Zimmerman, use similar devices.

[7] Here is a question about just what should be on the list. The sentence "George <u>finishes</u> climbing the tree at a moment earlier than this time" has a truth-value only relative to times; but the word "this" does not seem to have anything special to do with time. Nor is "this" on any traditional lists of A-vocabulary. The best solution I think is to nevertheless consider "this," along with other demonstratives, including "that," a piece of A-vocabulary.

containing it from having time-independent truth-values, if the block universe theory is true. Unfortunately this is not that helpful, since if we want to know which pieces of vocabulary meet this condition we will just get the list that appears in (2).

Statement (2) is still not quite right as a more precise statement of reductionism about tense. But the statement I want to use is close by:

(3) A complete description of reality can be given without using tense or A-vocabulary; moreover, a description meeting these conditions is more fundamental than one that does not.[8]

I have said something about tense and A-vocabulary, but I still need to say what it is for a description to be complete, and for a description to be fundamental. As a first pass we can say that a description is complete if and only if all truths follow from that description. But this is only a first pass. Look at it in more detail:

Let D be a collection of sentences that constitute a complete description, and S any true sentence. Then, necessarily, if all the sentences in D are true, S is also true.

If this were true then (3), and therefore the block universe theory, would be very easy to refute. Let D contain the tenseless sentences

- George <u>finishes</u> climbing the tree at noon on May 2, 2012.
- For each time T later than or identical to 11:55am, and earlier than noon of May 2, 2012, George <u>is</u> climbing the tree at T.
- For each time T earlier than 11:55am and later than noon, George <u>is</u> not climbing the tree at T.

D certainly seems like it is complete, at least where George's climbing habits are concerned. But now suppose that today is May 3, 2012. Then "George climbed yesterday" is true, but its truth does not follow from the truth of the tenseless sentences in D.

So in what sense does the block universe theory say that a tenseless and A-vocabulary-free description can be complete? Start with a claim we have already seen:

(4) Sentences (sentence types) with tense or A-vocabulary are true or false only relative to a time, and a typical sentence of that kind is true at some

---

[8] This is reductionism about tense broadly construed. It is the conjunction of reductionism about A-vocabulary, and reductionism about tense narrowly construed. For simplicity I usually use "reductionism about tense" for the broader thesis, but in Chapter 4 I will sometimes use it for the narrower notion.

times and false at others. "1865 is present," for example, is true only at times between the first and last instant of 1865. And "It is raining" is false now but true, so the weatherman tells me, at a time three days hence. Tenseless and A-vocabulary-free sentences, by contrast, have "time-independent" truth-values.

As I said earlier, while claim (4) may seem too obvious to be worth stating, we will later see alternatives to the block universe theory on which it is false. (I should mention that while the truth-values of tenseless and A-vocabulary-free sentences are independent of time they will usually depend on other features of the context in which those sentences are used, if they include words like "I" or "over there." But I will for the most part ignore kinds of context-dependence other than dependence on time, and sometimes say that a true sentence that has its truth-value independently of time is "just plain" true.)

The simpleminded way of understanding "complete description" demanded that a complete description of reality fix which sentences in a language containing tense or A-vocabulary are just plain true; but in light of (4) this demand does not make any sense.

Statement (4) is the first part of the explanation of what the reductionist means by "complete description." The second part is the claim that

(5) Every sentence containing tense or A-vocabulary has tenseless and A-vocabulary-free truth-conditions.

A truth-condition for a sentence S is a condition under which S is true. I admit that this is not very informative. It is easier to say something informative about what a "reductive" truth-condition, a truth-condition in a more fundamental language, is.

But before I can say what a reductive truth-condition is I need to say something about fundamentality. What does it mean to say that one language is more fundamental than another? This is a difficult philosophical question. I think we can get by without a detailed answer to it.[9] What matters for my purposes is that there is such a thing as fundamentality; it does not matter what its nature is. Of course, plenty of philosophers deny the existence of fundamentality. This is not the place to argue for its existence. That is just something I will assume in this book.

---

[9] A recent comprehensive discussion of this question is in Ted Sider's book *Writing the Book of the World*.

What I will do is say a little bit more to help you grasp what I am talking about (assuming you are not a fundamentality-denier, so that you do not already deny that there is something to grasp). Here is a true sentence: "The Statue of Liberty is green." But this sentence is not a fundamental truth; anyway, I do not think so. Color vocabulary does not belong to the fundamental language, and color talk does not get at the "ultimate nature of reality." Why not? One reason is that color categories lack unity. They do not "carve nature at its joints." Things can be the same color without really being very similar. If this is right, what more fundamental vocabulary should replace color vocabulary? One view is that reflectance-talk—talk of how things are disposed to reflect light—is more fundamental than, and should replace, color-talk.

Now the claim that reflectance-talk should replace color-talk is the claim that color-talk should be given reductive truth-conditions using reflectance-talk. This brings us back to what a reductive truth-condition is. A reductive truth-condition for a sentence S says in a more fundamental language what things have to be like in order for S to be true. Furthermore, if S is true, then it is true because things are the way specified in the reductive truth-condition. If the reductive truth-condition for "The Statue of Liberty is green" is "The Statue of Liberty is disposed to reflect light in such-and-such ways" then, since the Statue is green, it is green because it is disposed to reflect light in those ways.

Let us get back to the case at hand. The sentences that are "to be reduced," namely sentences containing tensed verbs and A-vocabulary, have their truth-values relative to times. The sentences "to which they are reduced" do not. So a tenseless and A-vocabulary-free truth-condition for a sentence that is "to be reduced" will specify the conditions under which that sentence is true at a time T. That is, a tenseless and A-vocabulary-free truth-condition puts a name for a sentence that is "to be reduced" in for "S" in

(6) For S to be true at T is for it to be the case that . . .

and fills in the right-hand side with a sentence that lacks tense and A-vocabulary.
Now any instance of (6) will entail an instance of

(7) Necessarily, S is true at T if and only if . . .

But not conversely; in "Necessarily, X if and only if Y" there is symmetry between X and Y, while in "For it to be that X is for it to be that Y" there is not. The second implies that Y is more fundamental than X; the first does not. Still, just for convenience I will usually write truth-conditions as instances of (7) (and omit the word "necessarily"); but it is always the corresponding instance of the strong form (6) that I intend.

Here is an example of a tenseless and A-vocabulary-free truth-condition for a tensed sentence:

(8) "A and B will be farther apart" <u>is</u> true at T if and only if there <u>is</u> a time later than T such that A and B <u>are</u> farther apart at that time than they <u>are</u> at T.

Now let C be the collection of all sentences that give the truth-conditions for sentences in the "to be reduced" language. Then reductionists say that a tenseless and A-vocabulary-free description D can be complete in this sense:

(9) Let T be any time and S any sentence that is true at T. Then necessarily, if all the sentences in D are true, and all the sentences in C are true, S is true at T.[10]

To see all this in action look back at the picture in Figure 2.7. The content of that picture can be "translated" into tenseless and A-vocabulary-free language (by sentences like, "A <u>occupies</u> spacetime points a, b, c, . . . and B <u>occupies</u> spacetime points x, y, z, . . ."). The tensed sentence "A and B will be farther apart" is a tensed sentence that is true at T1. Clearly that this sentence is true at T1 follows from the tenseless and A-vocabulary-free description the picture embodies, plus the sentence's truth-condition.

The sentence (8) is just an example, a truth-condition for one tensed sentence. Giving reductionist truth-conditions for all tensed language is a complicated enterprise. There are compound tenses like the past perfect ("He had climbed the tree") and phenomena like aspect ("She runs the race" and "She is running the race" are both in the present tense but differ in aspect).[11] And it would be bad to just write a long, long list of truth-conditions. Truth-conditions should be assigned systematically, making clear, for example, how the truth-condition for "George climbed" is related to the truth-condition for "George is climbing." Even if we ignore these complications, and also ignore sources of context-dependence besides tense, the truth-conditions for a simple tensed sentence may depend in complicated ways on features of the context besides the time, such as the history of the conversation in which it is used. If I ask in 2012 what things will be like between you and me next year and you reply "You and I will be farther apart" then the truth of what you say requires not just that you and I be farther apart at some time later than 2012, but that you and I be farther apart at some time more than a year but less than two years later than 2012. Having said this I will do very

---

[10] Strictly speaking, the "all the sentences in C are true" clause is not required, since C by definition contains only necessary truths. But putting it in makes it clearer what is going on.

[11] Ludlow surveys reductionists' attempts to handle these and other phenomena in *Semantics, Tense, and Time*.

little to defend reductionism about tense in this book. I will discuss the most well-known objection to reductionism about tense later, in the first appendix to Chapter 12.

I have said that associated with the block universe "view" is a picture and a theory. The picture presupposed the 4D view. But the theory does not. It is compatible also with the 3 + 1 view. True, the reductionist truth-conditions I have been using as examples presuppose the existence of times. But those times need not be regions of spacetime.[12]

2.3 The Denial of Passage

If the block universe theory is true, does time pass? Here are two simple-minded responses:

Of course time passes. Look at the clock—the second hand is moving!

Of course time passes. Look at the clock—it reads 10, while just a little while ago it read 8!

Defenders of the block universe theory agree that the second hand is moving, and that the clock has changed which number it indicates. What this change comes

Figure 2.8 The motion of a clock hand in the block universe.

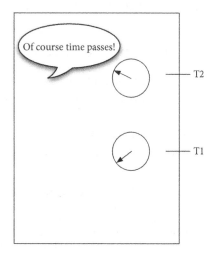

---

[12] The 4D view is the one that is correct; and since the actual spacetime geometry does not come with a preferred way of slicing it into a stack of instantaneous spaces, we will eventually need reductionist truth-conditions that do not presuppose the existence of such a preferred slicing. I will say something brief about this in the appendix to Chapter 9.

to, on the block universe theory, is depicted in Figure 2.8. The clock <u>reads</u> 10 at
T2 (which is the time at which this conversation occurs) and it <u>reads</u> 8 at T1 (the
time two hours earlier).

So the block universe view is certainly compatible with the claim that clocks
indicate different numbers at different times. If that is all it takes for time to pass
then time passes in the block universe.[13]

But many philosophers have thought that the passage of time is missing from
the block universe. They think that the passage of time requires something more
than the block universe gives us.

I said in the introduction that trying to figure out whether, if the block universe
view is true, there is any process that deserves to be called "the passage of time"
is a waste of energy. We should say that there is "anemic" passage of time in
the block universe but no "robust" passage of time. When C. D. Broad[14] tried to
describe what robust passage of time might be, he called it "Absolute Becoming."
I prefer the term "objective becoming," so I will sometimes use it as another name
for robust passage of time.[15]

As I have said, I believe the block universe theory. I deny the existence of any
robust passage of time. I deny that objective becoming occurs.

But this denial does not amount to much until we know what objective becoming would be if there were such a thing. In the next two chapters I will sift through
several theories of time that have been alleged to contain objective becoming to
find those that best capture what objective becoming is supposed to be.

2.4 Appendix: What the Block Universe Theory is Not

Reductionism about tense is a key part of the block universe theory. But the block
universe theory, as I intend it to be understood, need not be a more global form
of reductionism.

The way I have been describing the theory may encourage a more global
kind of reductionism. I said that in the theory "spatiotemporal reality is nothing

---

[13] Savitt, in "On Absolute Becoming and the Myth of Passage," favors a "deflated" understanding
of the passage of time like this. See also Dorato, "Absolute becoming, relational becoming and the
arrow of time," and Dieks, "Becoming, relativity and locality."

[14] Broad, "Examination of Mctaggart's Philosophy," p. 281.

[15] Sometimes a theory with robust passage is called an A-theory of time. Then the debate between
the block universe and robust passage is the debate between the B-theory and the A-theory. But this
is not good terminology. It is hard for non-experts to remember whether it is the A-theory or the B-
theory that has robust passage. Plus "A-theory" is too vague. Different philosophers use it to cover
different clusters of claims, which creates lots of opportunities for people to talk past each other.
But we need not be bound by the terminological mistakes of our forebearers. I will never write
"A-theory" or "B-theory" again.

but a four-dimensional block universe." Draw a picture of spacetime; indicate where in spacetime each spatiotemporal thing is; say what each of those things is like during each part of its career; then you have a complete picture of spatiotemporal reality.

I keep sticking the word "spatiotemporal" in there. Why? If the picture is a complete picture of spatiotemporal reality, isn't it also a complete picture of reality-without-qualification? If the theory does not leave out the passage of time (either because it is in the picture already, or because there is no such thing to leave out in the first place), what else could it be leaving out?

There are plenty of things. They constitute a list of topics of philosophical controversy. The picture may leave out what the laws of nature are; it may leave out which events are causes of which other events; it may leave out whether anything is conscious, who is evil and who is good, or which pieces of contemporary art are beautiful. When I say that a picture of the block universe is a complete picture of spatiotemporal reality I mean to remain silent about whether the picture leaves any of these things out. (The terminology is not entirely perfect, since nonreductionists about causation still think that causal facts are part of spatiotemporal reality.)

It is an interesting question whether spacetime and its contents really are all there is, in the sense that all facts are determined by, or grounded in, facts about them. I admit to feeling the pull of the yes answer sometimes. Where could these other facts come from? If global reductionism is false then the "further facts" about causation or about beauty float free of the world of rocks and trees and electrons and protons. Like ghosts and spirits they are by turns spooky and fascinating. Anyway, once again, I intend the block universe theory to be a reductive theory of tense and passage; not necessarily of anything else.

2.5 Appendix: Other Versions of Reductionism about Tense

Early reductionists attempted to write down tenseless and A-vocabulary-free sentences that were synonymous with tensed sentences. One early version of reductionism (there were others) said that "George will climb the tree" is synonymous with "George's climbing of the tree is later than this utterance" (see for example chapter 7 of J. J. C. Smart's *Philosophy and Scientific Realism*). This form of reductionism about tense is false. A. N. Prior gave this example: "Eventually all speech will have come to an end" is a tensed sentence that may well be true. But Smart's translation scheme says it is synonymous with "The end of all utterances is earlier than some utterance later than this one," which cannot possibly be true (*Past, Present and Future*, p. 12). Later reductionists abandoned the claim that tensed sentences are synonymous with tenseless ones and replaced it with the supposedly weaker claim that tensed language can be given tenseless and A-vocabulary-free truth-conditions.

Later reductionists, like Mellor in *Real Time* and *Real Time II*, also thought that reductionism should provide truth-conditions for sentence tokens, rather than for sentence types. I think this is a mistake. Reductionism as I formulated it speaks of truth-conditions for sentence types.[16]

There is some confusion surrounding the move from providing synonymous translations to "merely" providing truth-conditions that could be cleared up if we stated reductionism about tense using proposition-talk. Two sentences are synonymous, intuitively, if and only if they mean the same thing. Now propositions are supposed to be the things sentences mean; at least, this is a very common view. It follows that two sentences are synonymous if and only if they express the same proposition. But language is context-dependent. A sentence may express different propositions in different contexts. So the claim that tensed sentences are synonymous with tenseless ones (for now on I will ignore A-vocabulary in this appendix for the sake of brevity) has two readings:

(10) For any tensed sentence S there is a tenseless sentence T such that in any context C, S and T express the same proposition.

(11) For any tensed sentence S and any context C there is a tenseless sentence T such that S and T express the same proposition in C.

Statement (10) is certainly false, but (11) may well be true. I do not think that the discussions of early reductionism refuted it.

(The thesis that sentence-meanings are propositions is less common than it used to be. Since the work of David Kaplan, especially "Demonstratives," it is more common, at least among philosophers of language, to identify the meaning of a sentence with its "character." Characters are functions from contexts of use to propositions. On this view the claim that tensed sentences are synonymous with tenseless ones has just one reading, namely (10).)

Although (11) may well be true, I have not adopted it as my official formulation of reductionism about tense, for several reasons. First, using proposition-talk responsibly requires a clear conception of what role propositions are supposed to play in our philosophical theories. But there is some disagreement about what

---

[16] For more on the ins and outs of various forms of reductionism see: part 1 of Oaklander and Smith's *The New Theory of Time*; Paul, "Truth Conditions of Tensed Sentence Types"; and Torre, "Truth-conditions, truth-bearers and the New B-theory of time."

that role is, or even whether there is a single role here, rather than different roles that should be played by different things.[17] I want to make progress in the philosophy of time without having to stake out a view on propositions. Second, and relatedly, formulating reductionism about tense using proposition-talk in the way I just suggested makes it contentious in a way I would prefer to avoid. For example, some believers in the block universe hold that tensed sentences express propositions that can be true at some times and false at others. They hold that propositional truth is time-relative rather than absolute. These philosophers deny that a tenseless sentence can express the same proposition as a tensed one, even in a single context. But they still, I take it, want to endorse some form of reductionism about tense.[18]

---

[17] For discussion of this topic and arguments that no one thing can play all the roles traditionally associated with propositions see Yalcin, "Semantics and Metasemantics in the Context of Generative Grammar," and the references therein.

[18] See, for example, Brogaard, *Transient Truths*, for reasons why you might want to have a view like this.

3

What Might Robust Passage Be?

3.1 Passage and "Real Change"

If objective becoming requires more than the block universe provides, if it is not enough that a clock indicates 8 at one time and 10 at a later time, then what does it take for objective becoming to occur? One reason commonly given for saying that there is no passage of time in the block universe goes like this:

> Just look at the diagram (in Figure 2.8 for example, repeated here as Figure 3.1). There it is, all of spatiotemporal reality, spread out before you. *It does not change.* And if all of spatiotemporal reality is an unchanging four-dimensional block then there is no real passage of time.

Looking at this line of thought in more detail will lead us toward an account of what robust passage of time might be.

To defend this line of thought one needs to provide (i) an argument that nothing changes if the block universe theory is true, and (ii) an argument that this precludes the robust passage of time. The second argument is straightforward. All we need is the premise

> (1) If there is such a thing as the robust passage of time then something is changing (or has changed, or will change).

This premise looks fairly uncontroversial. It says only that change is a necessary condition for robust passage. It does not say that change is sufficient. And change is a necessary condition even for anemic passage.

(Turning (1) into a condition that is necessary and sufficient for the robust passage of time requires saying what must change for robust passage to occur. One natural idea is that time itself must change for robust passage to occur. But I want to postpone discussing this natural idea until later.)

So let us grant for now that (1) is true, that robust passage requires change. Why think that nothing changes in a block universe? Let's work up to an argument for this claim by looking at the opposition. Bertrand Russell, an early

Figure 3.1 The motion of a clock hand in the block universe.

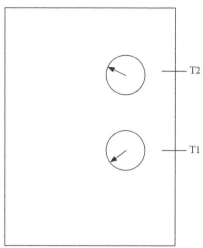

T2

T1

defender of the block universe, said that of course there is change in the block universe. He used his theory of change to defend this claim. Russell's theory of change starts with what looks like a platitude about change:[1]

(2) Something is changing if and only if (i) it is currently one way, and (ii) it was (not long ago) some other, incompatible way.[2]

The Russellian then combines this present-tensed statement with reductionism about tense to get

(3) Something _is_ changing at T if and only if it _is_ one way at T and some other, incompatible way at a nearby earlier time.

Now look again at the block universe with the ticking clock (Figure 3.1). The clock indicates 8 at T1 and 10 at T2. (Let us imagine that T1 and T2 are very close together.) Moreover, it seems that indicating 8 and indicating 10 are incompatible; nothing can do both. Given these facts, if (3) is true, the clock is changing at T2.

---

[1] See section 442 of Russell's _Principles of Mathematics_. Throughout this section I draw on the discussion in section 6.2 of Sider's _Four-Dimensionalism_.

[2] Statement (2) gives conditions under which something is (presently) changing. Similar platitudes state the conditions under which something will change and the conditions under which something has changed.

Two other points. First, (2) is correct only if "not long ago" means something like "at the previous moment in time." I say "something like" because time is dense; there is no previous moment. Accounting for this fact would require the use of calculus, which I prefer to avoid for now. So I will stick to the sloppier formulation in the text. Also, it may be that the definition should include as a third clause "(iii) it will be (shortly) some other, incompatible way." Someone might say that (i) and (ii) alone are compatible with the possibility that the thing has just stopped changing, that this is the first moment at which it is not changing. For simplicity I am going to ignore these complications.

What might one think is wrong with Russell's account? There are two places to attack it. Maybe Russell was wrong to accept (2). Maybe (2) does not give the conditions under which something changes. Or maybe Russell was wrong when he said that if (2) is true, and reductionism about tense is true, then the clock changes.

One well-known response to Russell is originally due to J. M. E. McTaggart. One way to put the idea driving McTaggart's conviction that there is no change in the block universe goes like this:[3]

> Yes, the clock indicates 8 at T1, but it always indicates 8 at T1. Similarly it always indicates 10 at T2. So there is no change.

One way of fleshing this out makes it an instance of the second kind of response to Russell. Even supposing that (2) is correct, the clock does not change.

The key to understanding this response is to focus on the word "way" as it appears in (2) and (3). Here is (3) again:

(3) Something is changing at T if and only if it is one way at T and some other, incompatible way at a nearby earlier time.

It might seem obvious that the clock in question is one way at T1 and another way at T2. But strictly speaking this only follows if we combine (3) with some claims about what ways there are for the clock to be. Another term for ways the clock can be is "properties the clock may have." Russell's argument assumes that among the properties the clock may have are these:

(P1) The property of indicating 8.
(P2) The property of indicating 10.

If it seemed obvious to you that, given (3), the clock changes, then you were probably also assuming that these are properties the clock may have. For if there do indeed exist such ways for the clock to be as (P1) and (P2), and if the clock does indeed have them (at different times), then it follows from (3) that the clock changes. The response to Russell we are now considering disputes this. It says that if the block universe theory is correct then there are no such properties as (P1) and (P2) for the clock to have. Instead the properties there are for the clock to have include

---

[3] What follows is a paraphrase of one of McTaggart's arguments against (2). (Both are in *The Nature of Existence*, volume 2, chapter 33, sections 315–16.) His other argument is that (2) makes change too analogous to spatial variation. A pencil is sharp at one place and soft at another; but this spatial variation does not constitute change. Similarly, the argument goes, "mere" temporal variation does not constitute change. This argument does not connect with the line of thought I am pursuing.

(P3) The property of indicating 8 at T1.

(P4) The property of indicating 10 at T2.

and, for good measure,

(P5) The property of indicating 8 at T2.

(P6) The property of indicating 10 at T1.

though the clock does not have (P5) or (P6) in the diagram. Now if there are no such properties as (P1) and (P2) then the clock does not first have one of them and then the other. If the only relevant properties are properties like (P3) and (P4) then the clock does not have one of them at one time and another, incompatible one at a nearby time. Instead it always has both (P3) and (P4). (These properties are not incompatible.) So if there are no other relevant properties then even if (3) is correct the clock does not change.

(Sentences like "The clock has, at T2, the property of indicating 8 at T1" and "The clock always has the property of indicating 8 at T1" may sound weird—though the first seems weirder than the second. The "always" and "at T2" seem like they should not be there at all. It sounds better to say "The clock has the property of indicating 8 at T1 full-stop," or "the clock just plain has" that property. Or at least, that is how things seem to those steeped in the block universe theory. We will later meet a theory in which these allegedly weird sentences are perfectly in order.)

Now where does the claim that properties (P1) and (P2) do not exist, while (P3) and (P4) do, come from? One principle from which this claim can be deduced (when combined with the block universe theory) is this one:

(4) There is such a property as the property of A-ing if and only if all sentences of the form "X As" are just plain true or just plain false. Consequently, if some sentence of the form "X As" has a truth-value only relative to a time then there is no such property as the property of A-ing.

Reductionists about tense say that for ordinary, working clocks, "The clock indicates 8" is true at some times and false at others. It is not just plain true or just plain false. If (4) is correct, then, reductionism entails that there is no such property as the property of indicating 8.

Fine. That is an argument that there is no change in the block universe. Just to remind ourselves where we are, the conclusion of this argument combines with

(1) If there is such a thing as the robust passage of time then something changes.

to entail that there is no robust passage in the block universe.

So how good is this argument that nothing changes in the block universe? It might be tempting to reject (4) and say that (4) does not correctly state the conditions under which properties exist. But this objection does not get at the heart of the argument. The argument can be reformulated to avoid it. Since this reformulation is better in several ways, I want to present it.

What might an alternative to (4) be? One idea is

> (5) There is such a property as A-ing if and only if all sentences of the form "X As" are (grammatically) complete sentences.[4]

The sentence "The clock indicates 8" is certainly a grammatically complete sentence. And there appear to be no other obstacles (if the block universe theory is correct) to inferring from (5) that (P1) exists.

Okay, so suppose (4) does not correctly give the conditions under which properties exist. Instead it just selects a special subset of all the properties. Call them, for no very good reason, the E-properties. The argument we are considering can be reformulated so that it does not use (4) at all. The new version works by replacing (2) with a different premise about change. This new version of the argument says that "mere" temporal variation in properties is not sufficient for change. Instead it uses this analysis:

> (6) Something is changing if and only if (i) it currently has one E-property, and (ii) it had (not long ago) some other, incompatible E-property.

Ted Sider has a nice way of reformulating (6).[5] If the block universe theory is true then sentences like "The clock indicates 8" are grammatically complete. But in another sense they are incomplete. Say that a *metaphysically complete* sentence is one that has a truth-value absolutely, not relative to anything else. A metaphysically complete sentence does not need to be supplemented with any additional facts to determine a truth-value. In the block universe "The clock indicates 8" has a truth-value only relative to a time and so is not metaphysically complete. Another way to write (6), then, is:

> (7) Something is changing if and only if there is some metaphysically complete sentence S about it such that (i) S is true (absolutely), and (ii) S was (not long ago) false.[6]

---

[4] Russell's paradox about the property of non-self-instantiation shows that this cannot be right. But this problem with (5) does not have any bearing on its role in the current discussion.

[5] It is in chapter 11 of *Writing the Book of the World*. He draws on ideas presented by Timothy Williamson in, among other places, "Necessary Existents."

[6] Statements (6) and (7) are equivalent, given the existence of properties. Since I do not believe in properties I like (7) better. I should say that (7) is meant to capture all kinds of change, not just "intrinsic" change.

Statements (7) and (2) give conflicting criteria for change. To have names for them I will say that the criterion in (7) is necessary and sufficient for "robust" change while the criterion in (2) is necessary and sufficient for "anemic" change. Note that if something changes robustly then it automatically changes anemically as well. When something changes anemically but not robustly I will say the change is "merely" anemic. With this terminology (7) is the claim that change, real change, is robust change. Merely anemic change is not enough.

The second way to run the argument that the block universe lacks robust passage, then, is to start with (7), infer that nothing changes in the block universe, and combine this claim with (1) to conclude that time does not robustly pass.

Now I do not think this argument is any good, because I reject (7). But my goal right now is not to evaluate the argument. My goal is to investigate what robust passage of time is supposed to be and what alternative theories of the nature of spatiotemporal reality contain that robust passage. The idea behind the argument I have been discussing is that robust passage requires robust change. I will later challenge this idea, but let us grant it for now. Can we go further? Is robust change not just necessary, but also sufficient, for robust passage? Probably the most popular alternative to the block universe theory is a theory called presentism. Presentism contains robust change. Does it contain robust passage?

3.2 Presentism

First I need to say what presentism is. Go back to the pictures associated with the block universe theory (one of which is reproduced in Figure 3.2).

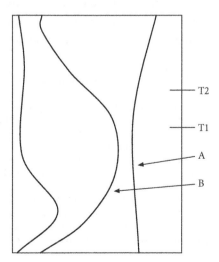

Figure 3.2 A picture of spacetime and its contents.

Many philosophers think that this picture, and the theory it accompanies, is completely wrong. One of their most famous opponents is A. N. Prior.[7] In one place he announces his opposition to the block universe like this:

So far ... as I have anything that you could call a philosophical creed, its first article is this: I believe in the reality of the distinction between past, present, and future. I believe that what we see as a progress of events *is* a progress of events, a *coming to pass* of one thing after another, and not just a timeless tapestry with everything stuck there for good and all. ("Some Free Thinking about Time," p. 47.)

The theory that spatiotemporal reality is a "timeless tapestry" is the block universe theory (though this is a misleading statement of it; the tapestry is not timeless, since it contains time as one of its dimensions). In another place Prior said this about why the block universe theory may be tempting:

It is tempting to think of the present as a region of the universe in which certain things happen, such as the war in Vietnam,[8] and the past and the future as other regions in which other things happen, such as the battle of Hastings and men going to Mars. ("The Notion of the Present," p. 246.)

Suppose that the block universe view is right, that Figure 3.2 is correct, and that it is now T1. Then there is such a region of reality (of "the universe") as the future. It is the region of spacetime above T1. Prior says here, no no no no. There is no such region. Nor, he says, does the past exist. There is only the present. This is presentism.

What exactly is presentism? What picture should we associate with it? Rewind to the picture that goes with the 3 + 1 view (Figure 3.3). Now remove time from the picture to arrive at Figure 3.4.

Take Figure 3.4 and fill it with material things and you will have a diagram that comes close to capturing presentism.[9] But wait—isn't it ridiculous to deny the existence of time?[10] Doesn't that denial have tons of absurd consequences? For

---

[7] It has probably been said before, but I will say it again. Prior had one of the best possible names for a philosopher of time.

[8] Prior delivered the paper from which this quotation is taken in 1969.

[9] Prior himself probably did not believe in the reality of space either. But if so his irrealism about space takes a very different form from his irrealism about time. In this context it is harmless to assume that space exists.

[10] Some presentists say that they do believe in the existence of times. (See, for example, Markosian, "A Defense of Presentism," and Bourne, *A Future for Presentism*. This kind of presentism goes back to Prior; he discusses it in various places, among others in "Tense Logic and the Logic of Earlier and Later.") But the times they believe in are "surrogates" for the times that exist in the block universe. Unlike the times in the 4D view these surrogate times are not regions of spacetime; unlike the times in the 3 + 1 view these surrogate times are not simple things (things without parts) that belong to a special "ontological category." Instead the surrogate times are certain sets of

Figure 3.3 Space and time on the 3 + 1 view.

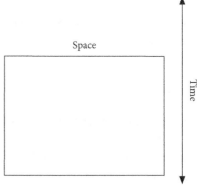

Figure 3.4 Spatiotemporal reality according to presentism.

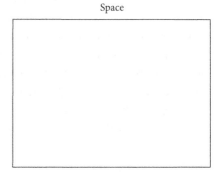

example, to move is to be in different places at different times. So if there is no time, motion is impossible.

Not so. Although presentists deny the existence of time they do not deny that our world is a temporal one. They are not mystics who say that temporality is a complete illusion. What they deny is that reality can have a "temporal aspect" to it only if there are such (strange!) things as instants of time.[11]

The temptation to think that presentism's denial of the existence of time is absurd comes from assuming reductionism about tense. Certainly if there are no times then it cannot be true that some given thing is in one place at the present time and a different place at an earlier time. Only if reductionism about tense is

sentences (or something similar)—what we would ordinarily call maximally complete descriptions of the universe-at-an-instant. To keep things simple I will ignore this kind of presentism. (Maybe presentism with surrogate times is a better theory than presentism without them, but not in any ways that are relevant to my topic.)

[11] Prior: instants, if they exist, are "a very odd kind of entity" ("Tense Logic and the Logic of Earlier and Later," p. 120).

true does it follow that if there are no times then nothing can move. But presentists reject reductionism about tense. They deny that to move is to be in different places at different times. Instead a presentist analysis of motion is in tensed terms:

Something is moving if and only if it is in one place and was (not long ago) somewhere else.

This tensed claim can be true, presentists say, even if there are no instants of time.

Although presentists are not reductionists about tense, those who accept the "standard" version (Prior's version) of presentism think that the way tense functions in English is not easy to see from the "surface form" of the language. In the surface form tense manifests itself in the variety of verb forms. Prior says that it is easier to see what is "really going on" if we regiment these sentences so that all verbs are in the present tense and non-present tenses are got by putting those present-tensed verbs inside the scope of tense operators. In general, a sentential operator is a piece of language that you attach to a sentence to form another sentence. Tense operators are special kinds of sentential operators. So Prior regiments "simple" past-tensed sentences, for example "George climbed the tree," by putting the tense operator "It was be the case that . . ." in front of the present-tensed sentence "George is climbing the tree" to obtain "It was the case that (George is climbing the tree)." For simple "future-tensed" sentences use the tense operator "It will be the case that . . ." More complicated sentences may require many nested tense operators.[12]

Because presentists are anti-reductionists about tense we have to interpret diagrams that go with their view differently from the way we interpreted diagrams that went with the block universe view. Block universe diagrams, like the one in Figure 3.2, correspond to tenseless descriptions of spatiotemporal reality.

---

[12] The transformations Prior makes do not result in well-formed sentences of English. "It was the case that George is climbing the tree" is ungrammatical. It would be a little better to write "It was the case that George climbed the tree." (In one place Prior likens this to emphatic double-negation, as in "I didn't see nothing.") A defender of Prior could say that his way of regimenting tensed sentences reveals the "logical form" of those sentences.

But nowadays linguists doubt that tensed sentences in natural languages like English are correctly represented as sentences containing tense operators and (only) present-tensed verbs. Partee's "Some Structural Analogies between Tense and Pronouns in English" is a canonical reference on this topic; see also King, "Tense, Modality, and Semantic Values."

Presentists can retreat and say that the language they favor is the correct one for describing how the world is fundamentally speaking, whether or not it correctly captures the semantic structure of English or any other natural language. If they say this they need to show how to give truth-conditions for sentences in natural languages like English in their preferred language. One strategy presentists might use to overcome the obstacles this project faces, and a defense of the idea that presentists might be able to get by giving something "weaker" than truth-conditions, may be found in Sider, "Presentism and Ontological Commitment."

Space

Figure 3.5 A presentist diagram.

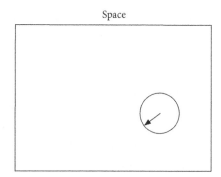

Figure 3.6 More presentist diagrams.

The way things will be in 2 hours...

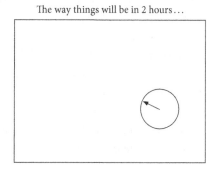

The way things are...

Presentist diagrams, like the one in Figure 3.5, do not. They correspond to tensed descriptions. So to interpret this diagram we need to know, is it a diagram of how spatiotemporal reality is (present tense)? Or a diagram of how it will be? Or a diagram of how it was?

A single presentist diagram, then, cannot be complete. A diagram of how things are leaves out how they will be. To have something complete we need a lot of diagrams—imagine extending the two in Figure 3.6 indefinitely upwards and downwards, as well as sticking some between the two that are there.

When you focus on Figure 3.4 there appears to be a clear difference between presentism and the block universe theory. But when you get to Figure 3.6 that appearance can start to fade. Now both theories seem to agree that spatiotemporal reality can be represented by a stack of three-dimensional spaces. The only difference seems to be that when I am talking about the block universe theory I put the entire stack into one diagram whereas when I am talking about presentism each space goes into a separate diagram. And it can start to look like these differences in how I choose to draw the diagrams do not correspond to any real differences in what the theories say.

I think that pictures and diagrams usually make things clear in a way words cannot, but perhaps in this case this is not true. What is hard to represent in the diagrams is the fact that the theories disagree about reductionism about tense. If you start to lose your grip on the difference between presentism and the block universe theory, reach for that difference, and then remember that reductionists need to believe in times for their truth-conditions for tensed sentences to be right, while anti-reductionists do not.

Whether presentism is true is a difficult question. Entire books have been written on it.[13] But I want to focus just on the question whether there is robust passage of time in the theory.

3.3 Passage and Presentism

I should say first that to a certain extent I think the question Is there robust passage in presentism? is a bad question. I do not think we should impose a litmus test that a theory must pass in order for it to contain robust passage. Presentism is a theory of time that departs from the block universe in some important ways but resembles it in others. Other theories of time I will discuss later also depart from it in some ways but not in others. They do not all depart in the same ways. I do not see much interest in legislating on which ways of departure make for robust passage and which do not. (They do have this in common: all of them reject reductionism about tense. But they do not all reject all of reductionism, or even the same parts.)

Still, having said all this, I think that the passage of time we get in Prior's version (which is the standard version) of presentism is less robust than the passage we get in the other theories I want to talk about. For this reason I will say little

---

[13] For example, Bourne's *A Future for Presentism*. Shorter but also relatively comprehensive treatments of presentism, with detailed bibliographies, are ch. 2 of Sider's *Four-Dimensionalism* and Markosian's "A Defense of Presentism."

about presentism in the rest of this book. (There is a version of presentism with more robust passage, but I think it is too weird to take seriously. I say a little bit about it in the appendix.)

Here is the line of thought that Prior pursues when he discusses what present-ism says about the passage of time.[14] (I have numbered the key element for later reference, and do not put all the points in exactly the way Prior does.)

- The process that constitutes the flow or passage of time is the change events undergo when they are first future, then present, then past. This process continues as these events recede further and further into the past. So, for example, time passes in part in virtue of the fact that Barack Obama's first election was in the future and now is in the past.
- But talk of events is misleading. Event-talk is useful, but fundamentally speaking there are no such things as events. The truth-condition in a funda-mental language for "The election was present and is now past" is a sentence about Barack Obama, not a sentence about some event called "the election." It contains tense operators and present-tensed verbs:

    It was the case that (Barack Obama is being elected) but it is not now the case that (Barack Obama is being elected).

- But this just amounts to saying that Barack Obama has changed. He has changed from someone who is being elected to someone who is not. So in this case at least talk about an event changing from being present to be-ing past is really just talk of an ordinary thing (or person, Barack Obama) changing in an ordinary way.

(8) Now although the claim, made of some particular thing, that it has changed implies an instance of

(9) It was the case that S, but is not now the case that S,

the converse is not quite true. (The reasons why need not detain us here; I say more about them in the appendix.) Nevertheless it is (9) that is important for us, for to say that time is passing is just to say that some instance of (9) is true.

So Prior comes close to identifying the passage of time with the existence of change. And, importantly, the change that goes on in presentism is the robust change that some opponents of the block universe are looking for. That is, it counts as change even by the demanding criteria set out in (7). For according

---

[14] He develops this line of thought most fully in "Changes in Events and Changes in Things."

to presentists, "Barack Obama is being elected" is a metaphysically complete sentence. It is not true at some times and false at others (for there are no times at which it can be true or false).[15] It is just plain false (and was just plain true).

I do not think that the passage of time as characterized by (8) is particularly robust. Imagine a really boring universe. For a simple example suppose that there is (and always has been, and always will be) just one material thing, a single electron, and it never moves or changes in any way. If presentism is true then in a scenario like this there is no sentence that was true but is false.[16] Every true sentence is always true, every false sentence is always false. So on Prior's view time does not pass. But I think that those who believe in robust passage will want to say that time could pass even in a scenario like this.

I do not mean to be arguing that robust passage of time should be compatible with the complete absence of change. I think that is certainly false. My argument is that the robust passage of time should be compatible with there being just one material thing, an unchanging electron. (Of course, on Prior's version of presentism, if that electron never changes, and it is the only material thing, then nothing changes.[17])

Another reason why (8) seems to me to be an anemic conception of passage is that it endorses this claim:

> In the single-electron scenario time would (always) have passed if the electron has moved just once.

For suppose it moved two years ago. Then it was the case that (it was the case 1 year ago that the electron is moving) but it is not now the case that (it was the case 1 year ago that the electron is moving).[18] And this is an instance of (9).

From one point of view the kind of passage of time that Prior locates in presentism is not that different from the anemic passage of time in the block universe. Remember the picture that illustrates the anemic passage that goes on in the block universe (Figure 3.7). It is the fact that the clock changes that constitutes the passage of time. Well this fact also constitutes the passage of time if presentism is true. For we have "It was the case that the clock reads 8, but it is not now the case that the clock reads 8." True, presentism associates a different picture, or set of

---

[15] Thus presentists reject claim (4) from Chapter 2—the claim that a tensed sentence may be true at some times and false at others.

[16] One might say, "I grant that the electron never changes, but maybe there are some events that change?" Not in Prior's version of presentism, since he denies the existence of events. In the appendix to this chapter I explore this question in a non-standard version of presentism.

[17] Actually, Prior was a theist, so he may have held that in addition to the changeless electron there exists a changing God. I mean to be discussing the atheistic version of Priorian presentism.

[18] This sentence uses the metric tense operator "It was the case 1 year ago that . . ." I will say more about these operators in Section 7.3.

Figure 3.7 The passage of time in the block universe.

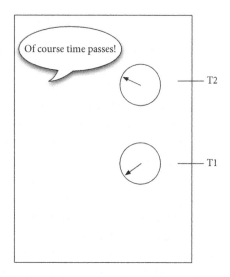

Figure 3.8 The passage of time in presentism.

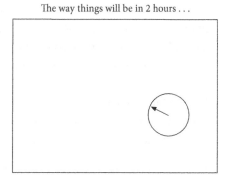

pictures, with a scenario containing a ticking clock. Instead of Figure 3.7 we have Figure 3.6, repeated here as Figure 3.8. But if our interest is in robust passage how big a difference is this really? True, in presentism this change is robust change while in the block universe it is not. That is a step toward most robust passage. But it is not a very big step.

3.4 Two Metaphors for Passage

I want to focus on theories that say that the robust passage of time requires more than that something change. What more? There are two natural thoughts. Perhaps the robust passage of time requires that time itself change in some way. Or perhaps the robust passage of time requires that some ordinary thing—something other than time itself, like a clock, or a monkey—change, not in some ordinary way, but in some extraordinary way (the ticking of a clock is merely an ordinary change).

These thoughts need development before we can look into whether the robust passage they aim to explicate really happens. With respect to the first thought, we need to know what kind of change time must undergo in order to robustly pass. With respect to the second thought, we need to know what kind of extraordinary change a clock, or a monkey, or a person must undergo for time to robustly pass.

The metaphors we commonly use to talk about the passage of time suggest answers to these questions. J. J. C. Smart articulated these metaphors as well as anyone in his paper "The River of Time."[19] Leaves and sticks may float down a stream. Sometimes, Smart says, we think that events are like leaves and sticks in the river of time. They float down the river of time from the future through the present into the past. Time is something that flows—moves—like a river flows (p. 483).[20] When we have this metaphor in mind we are thinking that the passage of time consists in time itself changing by (somehow) moving.

Smart articulates a second metaphor about time like this:

We say that we are advancing through time, from the past into the future, much as a ship advances through the sea into unknown waters. (P. 483.)

These two metaphors also appear, very briefly, in McTaggart:

we say that events come out of the future, while we [also] say that we ourselves move toward the future. (Sect. 306, note 1 of *The Nature of Existence*.)

And two years after Smart's paper D. C. Williams gives us the same two metaphors in almost the same language:

we may speak as if the perceiving mind were stationary while time flows by like a river, with the flotsam[21] of events upon it. . . . Sometimes, again, we speak as if the time sequence were a stationary plain or ocean on which we voyage. ("The Myth of Passage," p. 461.)

---

[19] Smart defends the block universe theory; he thought that these metaphors were "liable to lead us astray," that "when we talk of time as a river that flows we are talking in a way which is somehow illegitimate" (p. 484).

[20] Technically "Time flows like a river" is a simile, not a metaphor.

[21] My computer's dictionary (built-in to my Apple Macintosh) defines "flotsam" as "the wreckage of a ship or its cargo," or more generally "useless or discarded objects." I don't know what Williams

The second metaphor, the idea that the passage of time consists in our motion through time, is surprisingly (to me) common. I initially thought the first metaphor would be far more popular. Smart finds the second metaphor in Sir Arthur Eddington's book *Space, Time, and Gravitation*. Eddington wrote that "Events do not happen; they are just there and we come across them" (p. 51; quoted by Smart on p. 483). If we continue to liken time to a river, then the picture here is still one of events floating like leaves on a river. But instead of the river moving and carrying events along with it, we move ("advance") along the river of time and as we move we come across those events. For another example, the physicist Hermann Weyl wrote that

The objective world simply *is*, it does not *happen*. Only to the gaze of my consciousness crawling upward along the lifeline of my body, does a section of this world come to life as a fleeting image in space which continuously changes in time.[22]

Weyl seems to want to say that the passage of time is an illusion. But when he says what it is an illusion of he talks about crawling upward along his lifeline, which looks like a kind of motion through time.

A final example comes from Tim Maudlin. In his paper "On the Passing of Time" he also seems, at least sometimes, to identify the passage of time with our motion through time. He writes that when we ask about how fast time passes

we must mean to ask how the temporal state of things will have changed after a certain period of time has passed. In one hour's time, for example, how will my temporal position have changed? Clearly, I will be one hour further into the future, one hour closer to my death, and one hour further from my birth. So time does indeed pass at the rate of one hour per hour . . . (p. 112)

Maudlin's procedure for figuring out how fast time passes is to figure out how fast he moves into the future. This only makes sense if the passage of time consists, at least in part, in his, and presumably our, motion into the future.

So here we have two thoughts. The first is the thought that time itself changes by moving. The second is the thought that we change in some extraordinary way, by moving through time. These ideas are distinct. It might make sense to say that time itself moves without it making sense to say that we move through time. Do either of these ideas make sense? The next few chapters will be devoted to the idea that time itself moves. We will not get to the idea that we move through time until Chapter 10.

was trying to get at by likening events to flotsam. It strikes me as a nihilistic thing to say. I suspect that the metaphor of time as a river got out of his control.

[22] This quote is from *The Philosophy of Mathematics and Natural Science*, p. 116. I did not find it from reading Weyl but from reading other philosophers. It is quoted in both McCall, "Objective Time Flow" (p. 338) and van Inwagen, *Metaphysics* (p. 76).

3.5 Appendix: More on Passage and Presentism

I should start by saying more about the relationship between schema (9) and change:

(9) It was the case that S, but is not now the case that S.

I said under (8) above that the truth of an instance of (9) is not quite sufficient for there to be some particular thing that has changed. We must distinguish, as Prior did, between a *de dicto* and a *de re* reading of "something has changed." This sentence can mean "it was the case that (something or other is changing)," or it can mean "for some thing X it was the case that (X is changing)." The truth of an instance of (9) may be sufficient for the truth of "something has changed" on its *de dicto* reading, but—as Prior points out—is certainly not sufficient on its *de re* reading.[23] Consider the sentence

(10) The first election of George Washington is receding further and further into the past.

This sentence, like "The first election of Barack Obama was present and is now past," is supposed to express the idea that time is passing. Now the Priorian translation of (10) into a sentence that does not mention any events is, more or less,

(11) It was the case one year ago that (it was the case 224 years ago that George Washington is being elected) but it is not now the case that (it was the case 224 years ago that George Washington is being elected).[24]

---

[23] One kind of scenario in which the truth of an instance of (9) may not be sufficient even for the *de dicto* reading is one in which there have been instantaneous things. Suppose there was a statue that existed for just one instant, and that there were never any other statues. "It was the case that (there is a statue) but it is not now the case that (there is a statue)" is an instance of (9). Does this entail that it was the case that something is changing? Doubtful, because it is doubtful that an instantaneous thing can undergo change. (At least this is doubtful if presentism is true; we will later see a theory in which instantaneous things can change.)

Of course, even if the statue did not change, it could be that something else did. But what other candidates are there? If there is such a thing as the universe as a whole, then it could be the universe that was changing. It changed from having a statue-shaped part to not having a statue-shaped part.

One might of course deny that "all changes have a subject." That would be to deny that if there is change then something is changing. Then one might say that (9) is necessary and sufficient for it to be the case that there was change. (Prior introduces the term "quasi-change" and seems to use it the way we are here contemplating using "change.")

[24] This sentence contains metric tense operators, which, again I will not discuss in detail until Section 7.3. I say this is "more or less" the Priorian translation because (10), unlike (11), is silent about how far in the past the election is. The full truth-condition will be quite long but will contain (11) as one of its disjuncts. (One more quick point. One might think that the translation of (10) should include the claim that it was the case 225 years ago that Washington is being elected. ("His election used to be 224 years in the past, but it is now 225 years in the past.") But this follows from the first clause of (11) in standard metric tense logic.)

Statement (11) is an instance of (9). But (11) does not say of any particular thing that it has changed. For it to do that, (11) would have to be a statement about some particular thing; but it is not. The only candidate thing for it to be about is George Washington. But according to Prior's version of presentism there is no such person as George Washington. If there were he would exist "in the past," but there is no such place as the past for him to exist in. (Compare (9) to the analysis of change in (7); (7) adds to (9) the requirement that there be some individual that S is about.[25])

Prior says that a sentence like "George Washington is being elected," as it oc- curs in the sentence "It was the case that George Washington is being elected," means something like "Someone who chopped down a cherry tree and refused to lie about it and later led the American colonies' army against Great Britain is being elected." This is a sentence that is not about any individual in particular. (It is not about George Washington since it could be true if someone else had done all those things).

None of this matters very much. Prior does not identify the claim that time passes with the claim that there is some thing of which it is true to say that it has changed. He identifies the claim that time passes with the weaker condition that an instance of schema (9) is true.

I have argued that the conception of passage he ends up with is not very robust. How might Prior's presentism be modified to contain more robust passage? I am addressing this question in an appendix because answering it is a detour from the main thread of argument in the book, and taking the detour requires metaphysics more complicated than I want to do in the body of the book (this early in the body of the book anyway).

The first thing to say is that it will not help to adopt a version of presentism that affirms the existence of times. These versions of presentism say that times are certain kinds of abstract objects. One version identifies times with certain maximally consistent sets of tensed sentences. In more detail, the theory makes the following definitions:

T is a time $=_{df}$ T is a consistent set of tensed sentences; T is maximal (any tensed sentence consistent with T is a member of T); it either is, was, or will be the case that all and only the sentences in T are true.

Time T is present $=_{df}$ all and only the sentences in T are true.

Time T is later than time S $=_{df}$ T $\neq$ S and it either is, was, or will be the case that (S is present and T will be present).

---

[25] Statements (7) and (9) differ in a few other ways too, but they are not important in this context.

Now the theory can say that time is passing if and only if "presentness is moving into the future." More precisely,

(12) Time is passing if and only if some time later than the present time will be present.

This is all very nice, but is no improvement over Prior's account of passage in presentism. Let N be the present time and T a time later than N. Then "It will be the case that T is present" is true if and only if there is some sentence S such that

(13) S is now false, but will be true.

And to say that (13) is true is pretty much to say that some instance of (9) is true. The only difference is the use of the past tense in (9) where (13) has the present tense. But this difference makes no difference. Just as all instances of (9) are false in the single-electron scenario, in that scenario (13) is false for any value of S.

The fact that T is later than N only if T is distinct from N, and so contains a sentence S that N does not, plays a key role in this argument. I suppose a presentist might try to get (12) to capture robust passage by removing the requirement of distinctness from the definition of "later." Then the definition is just this:

Time T is later than time S $=_{df}$ it either is, was, or will be the case that (S is present and T will be present).

On this definition a time can be later than itself. Let us ignore the violence this does to the ordinary meaning of "later." If we adopt this definition of "later" then it will be true to say that in the single-electron scenario the present time is later than itself. So it will be true to say that a time later than the present (namely, the present time itself) will be present. Thus, by (12), in that scenario time is passing.

I really do not think this is an improvement. Even with the new definition of "later" on hand there is only one time in the single-electron scenario. That time is, always has been, and always will be present. So there is no change in which time is present. That does not sound like robust passage.

So let us once again set aside versions of presentism that affirm the existence of times. How else might one try to get robust passage into presentism? One might try to modify Prior's presentism by weakening his requirements for passage. Instead of (8), how about this?

(14) Time is passing if and only if, for some sentence S, it was the case that S is true.

This conception of passage is even less robust than Prior's. Prior's conception at least tries to capture the idea that the passage of time consists in "events receding further and further into the past." I do not think that Prior's conception captures this idea very well, but (14) does not capture it at all. For the condition in (14) is compatible with the possibility that events do not recede but instead stay fixed in time where they are. (More precisely, that condition is compatible with the "Priorian translation" of this statement about events into a statement that talks not of events but of ordinary things.) All that needs to be true for the right-hand side of (14) to be true is that the universe did not just begin to exist.

Is there a way get robust passage in presentism that keeps (8) and modifies some other part of Prior's theory? One part of his theory that does not seem essential to it is his denial of the (fundamental) existence of events. Might a version of presentism do better at containing robust passage of time if it embraced the existence of events?

I am with Prior in denying the existence of events, and I prefer to avoid discussing metaphysical theories that take the existence of events seriously. But let us see what taking it seriously can get the presentist.[26] Think again of the scenario with a single, never-changing, never-moving electron. Suppose the electron occupies point of space P, and consider the event E of the electron's being at P now. It is plausible to say that E is (presently) occurring, but it was not occurring until now and will not ever occur again. This entails that (switching to a more formal use of Priorian tensed language)

(15) It was the case that (it will be the case that (E is occurring)) but it is not now the case that (it will be the case that (E is occurring)).

So even though there is no sentence about the electron that, when put into schema (9), secures the passage of time, there is a sentence about E that does.

There is, however, a problem with this. For the following certainly seems true:

(16) It is never the case that there exists an event that is not occurring.

But now it looks like it cannot be right to say that "It was the case that (it will be the case that (E is occurring))" is true. Here is the argument (N is any positive number):

(17) It was the case N seconds ago that (E is not occurring).
(18) Therefore, it was the case N seconds ago that (E does not exist) [from (16), (17)]

---

[26] I will have to take events seriously again in Section 6.5.

(19) If it was the case N seconds ago that (it will be the case that (E is occurring)) then it was the case N seconds ago that (E exists).
Therefore, it is not the case that (it was the case N seconds ago that (it will be the case that (E is occurring))). [from (18), (19)]

It will not do to reject premises (17) or (19). Premise (19) follows from a much more general thesis, called serious presentism. Serious presentism asserts the truth of

- It is always the case that, for any thing X, it is always the case that (if X is F then X exists).

whenever "F" is replaced by a "positive" predicate.[27] The most plausible forms of presentism are serious. Premise (17) is just part of the description of the scenario.

So it seems that a presentist who endorses (8) and wants to use sentences about events to secure the passage of time in frozen worlds must reject (16). I think that a theory of events that rejects (16) is not just strange but crazy.

There is, however, a way around this argument. Claim (18) follows from (16) and (17) only with the unstated premise that whenever E exists it is an event. A presentist could deny this, and say that while E is currently an event, it is not always an event; E is an event only when it is occurring. This fits best with a version of presentism that says that nothing ever comes into or goes out of existence; everything always exists (and this is always true).[28] On this view George Washington, for example, continued to exist after he died. What happened when he died (and his body had decayed sufficiently) is that he ceased to be located in space, and ceased to have any other interesting intrinsic properties. Similarly,

---

[27] To get from serious presentism, with "is occurring" in for F, to (19), begin by dropping the initial "always" operator. This gives us

- For any thing X, it is always the case that (if X is occurring then X exists).

But if the embedded sentence is true of each thing it is true of E:

- It is always the case that (if E is occurring then E exists).

And if "if E is occurring then E exists" is always true then it was true N seconds ago:

- It was the case N seconds ago that (if E is occurring then E exists).

Finally, the operator "It was the case N seconds ago that" obeys the K axiom: if O(if X then Y) then (if OX, OY). So we get the desired result:

- If it was the case N seconds ago that (E is occurring), then it was the case N seconds ago that (E exists).

The predicates "X does not exist" and "X is not occurring" are not "positive." I do not have a useful characterization of positive predicates. A detailed discussion of the modal version of serious presentism may be found in section 4.1 of Williamson, *Modal Logic as Metaphysics*.

[28] Timothy Williamson defends a view like this in *Modal Logic as Metaphysics*. Sider in fact calls this version of presentism "Williamsonian passage" (*Writing the Book of the World*, p. 257).

on this view there is presently a thing that was George Washington's first election, and so was an event. But it currently is not an event. It currently has no interesting intrinsic properties.

On this view E has always existed, even though it only occurs now. Statement (15) is true, and so time passes. This is consistent with (16) because E was not an event until now.

So there is a version of presentism in which time passes even in worlds that are always frozen. I think it is weird enough to ignore, so I will ignore it. (I am not going to mount any arguments against it here; we have reached a point where, in this book at least, argument comes to an end.)

4

The Moving Spotlight

4.1 Can Time Itself Move or Flow?

If the block universe theory is correct then there is no sense to be made of the idea that time itself moves. Why not? One might answer by saying, "look down at the block; time, or spacetime, just sits there; it does not move." But this is a bad answer. Saying that time "just sits there" suggests that time is at rest. But in the block universe theory saying that time is at rest is just as much a mistake as saying that time moves. It makes no sense to ascribe either motion or rest to it.

One might explain why this is so by citing the fact that there is nothing in the block universe for time to be in motion, or at rest, with respect to. But while this is true it is not the thought that motivates me to reject the idea that time moves. Instead, the motivating thought is that spacetime is the "arena" in which motion takes place, and it is something like a category mistake to apply the concept of motion to the arena itself.

So the idea that time moves does not make sense in the block universe. Neither does it make sense if presentism is true. For if presentism is true there is no such thing as time to do the moving.

If there is a theory of spatiotemporal reality according to which time moves it must be a theory we have not yet met. Of the theories of spatiotemporal reality that philosophers have proposed, the one that comes closest to capturing the idea that time itself moves or "flows" is the moving spotlight theory. Let us explore this theory.

The first thing to say is that there is not really any such thing as "the" moving spotlight theory. There are many moving spotlight theories. But perhaps the most straightforward one works like this. Start with the block universe theory. To make things easy, use the picture of the block universe you get in the 3 + 1 view, but remove space from the picture so all we see is time. The result is Figure 4.1 (I have re-oriented time in the diagram from the vertical to the horizontal).

The moving spotlight theory says that this picture leaves something out. It leaves out which instant of time is present.

Time

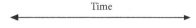

Figure 4.1 Time in the block universe.

Now it is a little tricky to say in what sense Figure 4.1 leaves out which time is present. For this is supposed to be a claim that believers in the block universe will reject. But I, who accept the block universe theory, agree that—in a certain sense—the picture leaves out which time is present. For I think that the following speech is true (it was produced in the year 2012):

If it is noon, New Years Day, 1965, then noon, New Years Day, 1965 is present. If it is midnight on the summer solstice in 1066, then midnight on the summer solstice in 1066 is present. In fact it is some time during 2012. But the picture does not indicate that some time during 2012 is present.

What I do disagree with is the claim that, regarded as a picture of how reality is at a fundamental level, Figure 4.1 leaves something out. And that is the claim that the moving spotlight theory makes. The moving spotlight theory rejects reductionism about A-vocabulary. It regards "which time is present?" as a question that uses only fundamental vocabulary and that has a definite answer.[1]

(Why not just say that, in the moving spotlight theory, the question "which time is present?" makes sense, and has an answer, "from a perspective outside of time"? That sets it apart from the block universe theory, which says that this question only makes sense relative to a time (or in a context that determines a time of utterance). Some versions of the moving spotlight theory do say this, but not all of them.)

What does a complete picture look like on the moving spotlight theory? The picture in Figure 4.2 is an improvement. The dot on the time line in the picture indicates which time is present.

But this picture also does not entirely capture the moving spotlight theory. For the moving spotlight theory does not just say that exactly one time is present. The theory also says that which time is present keeps changing. A picturesque way of putting this is to say that "presentness" "moves" along the series of

Figure 4.2 The present in the moving spotlight theory.

[1] When two theories disagree about the fundamental vocabulary it can be difficult to know how to characterize disputes between their proponents. I have said that believers in the block universe reject the claim, made by the moving spotlight theory, that exactly one time is present. Should this rejection consist in the claim that "T is present," as his opponent intends it to be understood, is false, no matter what goes in for T? Or should it consist in the claim that it is meaningless?

times from earlier times to later times. C. D. Broad introduced this picture in a well-known passage:

We are naturally tempted to regard the history of the world as existing eternally in a certain order of events. Along this, and in a fixed direction, we imagine the characteristic of presentness as moving, somewhat like the spot of light from a policeman's bull's-eye traversing the fronts of the houses in a street. What is illuminated is the present, what has been illuminated is the past, and what has not yet been illuminated is the future.[2]

But how can presentness move along the series of times? In the block universe ordinary material things move through space by being located in different places at different times. If that is what motion through space is, and if motion through time is a process strictly analogous to it, then presentness can move along the series of times only by being located at different times at . . . what? It seems that there needs to be some new "dimension of indices," distinct from time, so that different times can be present at, or relative to, or with respect to, different points in that dimension.

One version of the moving spotlight theory says that there is indeed such a dimension of indices. That dimension is a lot like time, but is still distinct from time. It is something new. I like to call it "supertime." (Sometimes it is called "hypertime.") This is The Moving Spotlight Theory with Supertime (for short: MST-Supertime). Supertime is like time in that it is a one-dimensional thing made up of points. And, the theory says, at each point of supertime exactly one instant of time is present. Later instants of time are present at "later" points of supertime, in such a way that "presentness" "moves" continuously.[3] The picture we get with MST-Supertime is in Figure 4.3; the arrows indicate which instant is present relative to a given point in supertime.

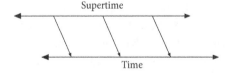

Figure 4.3 The moving spotlight with supertime.

---

[2] This quotation is from *Scientific Thought* (p. 59). A similar passage appears in Broad's *Examination of McTaggart's Philosophy*, volume 2, part 1 (p. 277). Broad's picture is a little different from the one I am using. He imagines presentness moving along a series of events rather than a series of times.

[3] I put the second "later" in quotes because someone might think that the word applies only to pairs of times, never to pairs of supertimes. Such a person would regard my use of the word here as deviant. If such a view is right we will need a new word to talk about how the points of supertime are ordered. I think it easiest to just extend our use of "later" to cover both order in time and order in supertime, so I will not put my future uses of "later" in scare-quotes.

Seeing Figure 4.3 lets us see in what way Figure 4.2 is incomplete. Figure 4.2 just shows what temporal reality is like from the "perspective" of one point in supertime. But from the perspectives of other points in supertime other times are present.

Now the idea that there is such a thing as supertime is crazy. It is just insane. But forget about that for a minute. I said that, of the theories of time that have been proposed, the moving spotlight theory is the one that comes closest to capturing the idea that time itself moves or flows. So does time move in MST-Supertime?

It does not. To see why let's think again about ordinary motion through space. When something moves through space there are three things involved. First there is the thing that moves—a plane, maybe, or a train, or an automobile. Second there is the dimension in which it moves; in the case of ordinary motion that is space. And third there is the dimension "with respect to which" it moves; in the case of ordinary motion that is time.[4] Now if time moves it certainly does not move through space with respect to time. But we still need three things: the thing moving, the dimension in which it moves, and the dimension with respect to which it moves. And there are not three things in Figure 4.3. There are only two. We have time; it is the thing that is supposed to be doing the moving. We have supertime; presumably it is the thing with respect to which time is moving. But what is the dimension in which time moves? There does not seem to be one.[5]

If anything is moving in the moving spotlight theory it is, obviously, the "spotlight," namely presentness, the property of being present. Now I myself am a

[4] "What do you mean 'in the case of ordinary motion' it is time? Isn't motion of any kind necessarily with respect to time?" I do not think so. Sometimes we do talk about something moving with respect to a dimension other than time. A picture of a sine wave moves up and down. That is, relative to different points along the x-axis (a spatial dimension) it is at different heights along the y-axis. I suppose that someone might say that this is not motion strictly speaking, that motion strictly speaking is always motion relative to time. But we do say things like "the sine wave moves," so there is such a thing as "motion loosely speaking" and it need not be relative to time.

[5] It is not really true that the ordinary motion of, say, a train, requires the existence of the train, and space, and time. A presentist will say that the train is moving if and only if the train is in one place and will be (shortly) in another place, where the truth of this tensed sentence does not require the existence of time. Fair enough. This does not affect the point I want to make. The point is that ordinary motion requires that there be "temporal variation" in something's position in space. So time itself can move only if we have things to put in for X and Y in the formula "There is X-variation in time's position in Y." The problem is that the moving spotlight theory does not have words to stick in for both X and Y. It does have something to stick in for X. The motion of time is "supertemporal" variation in time's position. But we still do not have anything to stick in for Y.

MST-Supertime says that supertemporal variation is variation with respect to supertime. But we can imagine a version of the theory that treats supertime the way presentism treats time. I will have a lot more to say about a version of the moving spotlight theory like this in Section 4.2.

nominalist; I do not believe in the existence of properties.[6] If I were to go for a version of the moving spotlight theory it would be a nominalist version. A nominalist version says that there is change in which time is present, but denies that the claim that there is such change is equivalent to the claim that presentness moves along the series of times.

I am not even sure that an anti-nominalist version should say that it is literally true that presentness moves along the series of times. Are properties really the sorts of things that can move? Some philosophers do say yes. But we are straying from the path, so let us set this issue aside.

To get a theory in which time itself moves we need to say that there is yet another dimension, other than space, time, and supertime. Then each time can move through this dimension by having a different location in it relative to different points of supertime. Here in more detail is how the theory might go. Let's use "B-time" to name the thing that I have so far been calling "time." And let's call the new one-dimensional thing "A-time." MST-Supertime said that each instant of (B-)time is present at some point in supertime. This new theory does not say this. It does not speak of times being present at, or with respect, to points of supertime. Instead it says that exactly one instant of A-time is present—and it is present full-stop, it is just plain present. In the terminology from Section 3.1 sentences like "T is present" are metaphysically complete.

But the new theory still has supertime in it. What varies from point of supertime to point of supertime are B-times' locations in A-time. At later and later points of supertime B-time has slid farther and farther along A-time. The picture that goes with this theory is the one in Figure 4.4. The first panel shows the

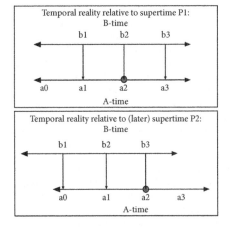

Figure 4.4 Time moving.

---

[6] I have used property-talk in this book, but I intend it only as a short-hand way for expressing claims that can be expressed without using property-talk.

relation between A- and B-time relative to one point in supertime P1; the second panel shows the relation relative to a later point P2. An arrow from a B-instant indicates the instant of A-time at which it is located, and the dot indicates which instant of A-time is present.

If we say—as we should—that A-times earlier than the present A-time are past A-times, and A-times later than the present A-time are future A-times, then in the picture each B-time moves from the future through the present into the past.

When McTaggart described what the passage of time would be if there were such a thing (which he denied) he distinguished between something he called "the A series" and something he called "the B series," and he said that the passage of time consists in "the B series . . . sliding along a fixed A series" or "the A series . . . sliding along a fixed B series." A-time and B-time are not quite what McTaggart had in mind when he defined "the A series" and "the B series," but I think that this theory captures the idea these quotations suggest.[7]

Even though it is literally correct to say that time moves in this new theory, it will not play any role in the rest of the book. That is because I think the theory is completely ridiculous. I just cannot bring myself to take it seriously as a theory of the world that might well be true. I am willing to believe in the existence of time, but not in the existence of both B-time and A-time.

I think that those who believe in objective becoming should give up on the idea that the robust passage of time consists in time itself literally moving. It should be enough to have a theory in which time as a whole changes. The moving spotlight theory is a theory like that. Let us return our attention to it.

4.2 The Moving Spotlight via Supertense

I think that MST-Supertime is the worst version of the moving spotlight theory. I have already complained that belief in supertime is just too much to ask. In the face of the proposal that there really is such a thing as supertime I can only stare incredulously.

That, anyway, is the source of my resistance to belief in supertime. I can give some reasons why no one should believe in supertime, though I have to admit that they are not the reasons that move me. (Nor are they reasons that would move me if I did believe in objective becoming.)

---

[7] The quotation is from *The Nature of Existence,* volume 2, section 306, note 1. It is closer to McTaggart's definitions to have "the A series" and "the B series" denote the same thing: the sequence of instants of times. When we speak of the A series we have in mind instants characterized as either past, present, or future, while when we speak of the B series we have in mind instants characterized as either earlier than or later than one another.

First, there are versions of the moving spotlight theory that do without super-time, and do not postulate any other new weird thing in its place. I will describe them below. If supertime is not needed to "get presentness moving" then it is extravagant to believe it exists.

A second reason worth mentioning goes like this:

The world according to MST-Supertime looks too much like a five-dimensional block universe. Look again at Figure 4.3. In that picture temporal reality as a whole does not change. The picture presupposes the 3 + 1 view, which treats space and time as distinct, but if we were to take the spacetime point of view and not ignore the spatial aspects of reality then the picture would be a picture of a five-dimensional manifold with two "time-like" dimensions (time and supertime). And presentness would be just a diagonal streak across the block. If a four-dimensional block universe lacks robust passage then so does a five-dimensional one.

Why think that a five-dimensional block universe lacks robust passage? One answer is that robust passage requires that time undergo robust change. In MST-Supertime each time is present at one point of supertime and non-present at the others—there is supertemporal variation in which time is present—but "T is present" is not metaphysically complete, so this is merely anemic change.

I think that believers in objective becoming should be allowed to disagree about whether objective becoming requires robust change or not. The (anemic) "movement" of presentness in MST-Supertime is still a kind of passage that is more robust than anything that goes on in the block universe. Maybe it is enough to account for the evidence that (supposedly) favors passage. (We will get to a discussion of that evidence in Chapters 11 and 12.) So while some friends of objective becoming may think that MST-Supertime does not adequately capture their view, others may think it does.

Another objection to MST-Supertime is that if there is such a thing as super-time then objective becoming requires not just that there be robust passage of time but also that there be robust passage of supertime. Broad and Smart both press this objection, and it has been repeated many times since. Here is Broad:

[supertime] is precisely like [time] in all those respects which led people to say that presentness "moves along" the [time] series. Such people [those who favor objective be-coming] ought therefore to say, if they want to be consistent, that presentness "moves along" [supertime] as well.[8] (*Examination of McTaggart's Philosophy*, volume 2, part 1, p. 279.)

---

[8] As I mentioned earlier, Broad actually talks about presentness moving along a series of events rather than a series of times. I have changed the terminology so the objection applies to the way I have characterized the moving spotlight theory.

And Smart puts the point like this (he is writing here not from his own point of view but from the point of view of those who believe in objective becoming):

just as we thought of the first time-dimension as a stream, so will we want to think of the second time-dimension as a stream also. ("The River of Time," p. 484.)

Yet there is no robust passage of supertime in MST-Supertime.

Broad says that consistency requires believers in MST-Supertime to say that supertime undergoes robust passage, but by "consistency" he cannot mean logical consistency. The theory as stated, with no robust passage of supertime, is logically consistent. Broad must mean some stronger sense of consistency, which requires one to treat similar things (time, supertime) in similar ways. But what exactly is bad about treating time and supertime differently?

There is an argument that MST-Supertime is defective if supertime does not undergo robust passage, but I do not see it in Broad or Smart.[9] It goes like this: robust passage of time requires that there be change in which time is present. But there is change in which time is present only if different times are present relative to different points in a time-like dimension. And—crucially—a dimension is time-like only if it undergoes robust passage. So there can be robust passage of time in MST-Supertime only if there is also robust passage of supertime.

Now, as we will see, not all fans of robust passage grant that change in which time is present requires supertime in the first place. Even among those that do, not all will grant the key premise that a dimension is time-like only if it undergoes robust passage. But if we do grant these premises MST-Supertime fails to contain robust passage.

MST-Supertime can be saved from this argument if supertime can be made to undergo robust passage. But now we are right back where Broad and Smart wanted us to be. After the passages just quoted, each philosopher writes that if the passage of time requires the existence of supertime then the passage of supertime requires the existence of a third time-like dimension ("superdupertime"?), and so on. The argument here is that if MST-Supertime is the right way to account for the robust passage of time, and if every time-like dimension must undergo robust passage, we will have to believe in an infinite number of time-like dimensions.[10] But I think that George Schlesinger showed this argument to be bad.[11]

---

[9] Brad Weslake brought the importance of this argument to my attention.

[10] The accusation that robust passage requires an infinite number of time-like dimensions also appears in many places besides Broad's and Smart's writings; for example, in Williams, "The Myth of Passage" (p. 464). (It is a good question why these authors think that believing in an infinite number of time-like dimensions is bad in a way that believing in two of them is not.)

[11] See *Aspects of Time*, p. 32.

The existence of supertime helps secure the robust passage of time because with it the claim "which time is present changes" can be understood as the claim that different times are present relative to different points in supertime. To secure the robust passage of supertime in the same way we need to understand "which supertime is superpresent changes" as the claim that different supertimes are superpresent relative to different points in some time-like dimension. But we do not need a third time-like dimension for this. Time itself can play that role. MST-Supertime can be augmented so that it says, not just that each time is present at some point in supertime, but also that each point in supertime is superpresent at some instant in time.

For a picture see Figure 4.5. Solid arrows indicate which time is present at a given point in supertime, and dotted arrows indicate which supertime is superpresent at a given instant in time. I do not like this theory at all, but the problem with it is not that there is no superpresentness moving along the series of supertimes.

Let's get back to the main goal of this section, namely finding an alternative version of the moving spotlight theory, one that does not postulate supertime. Broad thought that this could not be avoided: "If there is any sense in talking of presentness moving along a series of events [or times] . . . we must postulate a second time-dimension" relative to which it moves (*Examination of McTaggart's Philosophy,* volume 1, part 2, p. 278). But he was wrong. We have already seen a theory in which it makes sense to talk of trains moving along a track that does not postulate even a first time-dimension, namely presentism. So one way to do without supertime is to do to supertime what presentism does to time. The role talk of time plays in the block universe theory is played in (Priorian) presentism by tense operators. Let the role talk of supertime plays in MST-Supertime be played in the new theory by "super"tense operators. For reasons I will explain, I will prefix these tense operators with "super": "It super-is the case that . . . ," "It super-will be the case that . . . ," and "It super-was the case that . . ." This theory is the moving spotlight theory with primitive supertense (MST-Supertense). It says that exactly one time super-is present (full-stop), that each later time super-will be present, and that each earlier time super-was present. (This is not a complete statement of the theory. I will postpone giving a complete statement until Chapter 7, where I

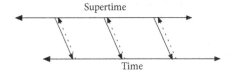

Supertime

Time

Figure 4.5 Schlesinger's moving spotlight theory.

discuss how fast time passes in the moving spotlight theory. I say a bit more about why the supertense operators are adequate replacements for talk of supertime in Section 7.2.)

Of course someone who accepts MST-Supertime can make sense of sentences containing these supertense operators. He will just say that the supertense operators are not fundamental pieces of vocabulary. According to MST-Supertime, the truth condition for, say, "It super-will be the case that T super-is present," is

"It super-will be the case that T super-is present" is true at supertime point P if and only if T <u>is</u> present at some supertime point later than P.

MST-Supertense rejects this truth condition, of course, and says that the supertense operators themselves are fundamental, and that sentences of the form "T super-is present" are metaphysically complete. They have their truth-values absolutely, not relative to points in supertime, or anything else.[12] Like presentism, and unlike the block universe theory, MST-Supertense is a version of anti-reductionism about tense (narrowly construed). (MST-Supertime, by contrast, rejects reductionism about A-vocabulary, but not about tense.)

As with presentism, no one picture captures MST-Supertense. We need a picture that shows which time super-is present and other pictures that show which times super-have been and which super-will. Figure 4.6 has two of them.

I will sometimes abbreviate sentences containing supertense operators using supertensed verbs. (That is, instead of writing "It super-was the case that George super-is climbing a tree" I will write "George super-climbed a tree.") Now supertensed verbs do not mean the same as the "ordinary" tensed verbs that occur in ordinary English. So how exactly do supertensed verbs work, and what is their relationship to tensed verbs in ordinary English? These are difficult questions for MST-Supertense. I will work toward what I think are the best answers.

Figure 4.6 Temporal reality in MST-Supertense.

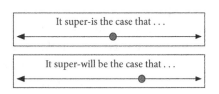

---

[12] The point about metaphysical completeness is most important. The supertense operators need not be fundamental. Parsons, for example, contemplates a version of the moving spotlight theory in which tense is analyzed using counterfactuals ("A-Theory for B-Theorists"). Bigelow, in "Worlds Enough for Time," and Schlesinger, in "The Stream of Time," also offer modal analyses of tense. (It nevertheless seems to me wrong to conflate temporality and modality as these theories do.)

Supertensed verbs are "tenseless with respect to time." So in one important respect they resemble the tenseless verbs used by the block universe theory. In that theory a fully explicit tenseless sentence about, say, the color of a square, must refer to a time. The block universe theory says that "The square is black at T" is complete, while "The square is black" is not.[13] The same goes for super-tensed sentences about the square's color in the language of MST-Supertense. "The square super-is black" is not complete. It super-is black at some times and not at others.

What is the relationship between supertensed language and "ordinary" tensed language? MST-Supertense, as I have said, rejects reductionism about tense. It takes supertensed language as basic. But it does not take all tensed language as basic. "Ordinary" tense, tense as it appears in ordinary language, does have an analysis in MST-Supertense.

Why is that? The main reason is that ordinary tensed talk is "one-dimensional." Roughly this means that if one wants to analyze ordinary tensed talk using quantification over instants of time, one needs only a single time dimension to do it. (That is exactly what happens in the block universe theory.) But the moving spotlight theory has two time dimensions. (Well, MST-Supertime has two. MST-Supertense has just one, but the shadow of the other is still there, in the supertense operators.) So the moving spotlight theory needs to say what in its two time dimensions our "ordinary" one-dimensional tensed talk is "tracking."

I think this is an important topic, but I should say that the material that follows is tough going. If you have no intrinsic interest in the question it addresses you may want to skim it, or skip it altogether and begin reading again with Section 4.3, returning here later if you find a need. (The only part of a later chapter that depends on this material appears in Section 9.4, and it is not terribly important. Part of Section 4.3 addresses a related question for another version of the moving spotlight theory.)

Figure 4.7 depicts a scenario that can serve as the focus for our discussion. In it a square super-is black at all times up to and including 2012. At later times it super-is white. Moreover, it super-will be, and super-was, those same colors at those times. We want to know whether the "ordinary" tensed sentence "The square will be white" super-is true or false in this scenario.

A natural thought to have is that this question makes no sense. "The square will be white" super-is true at some times and false at others; it does not super-have

---

[13] Of course, in a given context it may be clear how an utterance of "The square is black" is to be completed. On some occasions it might mean "The square is black at noon," on others it might mean "The square is black at some time or other," and so on.

Figure 4.7 Adventures of a square.

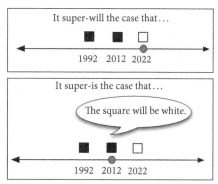

a time-independent truth-value. Suppose this is right. We can still ask whether "The square will be white" super-is true at times in 2012. I take it that the moving spotlight theorist will want to say that this sentence super-is true at times in 2012.

If this is right, then it rules out the most straightforward way of analyzing ordinary tense using the supertensed language of MST-Supertime. The most straightforward way is to follow the procedure "change ordinary tense to super-tense." This procedure generates the following truth-condition for our target sentence:

(1) "The square will be white" super-is true in 2012 if and only if the square super-will be white in 2012.

But the left-hand side of (1) is true and the right-hand side is false. (Even though the moving spotlight theorist should deny that the procedure "change ordinary tense to supertense" is the right one to use in general when providing truth-conditions for ordinary tensed language, she will say that it works in certain special cases. She should say that the truth-condition for "2050 will be present," where "will" here exhibits ordinary tense, is "2050 super-will be present." But let's set special cases like this aside.)

Another approach to providing truth-conditions for ordinary tensed language is to mimic the truth-conditions the block universe theory gives. The theory could take the block universe truth-conditions and change the tenseless verbs to verbs in the present supertense. This is the truth-condition we get for our example:

(2) "The square will be white" super-is true in 2012 if and only if the square super-is white at a time later than 2012.

Now the right-hand side is true. But this truth-condition is still no good. It divorces our ordinary tensed talk too far from the tensed language used to express the claim that there is robust passage of time. Statement (2), a truth-condition for a sentence in the "ordinary" future tense, contains no verbs in the future supertense.

I have looked at two ways one might try to analyze ordinary tensed language in MST-Supertense. The first had the ordinary future tense "track" what super-is happening "later in supertime." The second had the ordinary future tense track what super-is happening later in time. Each fails because it completely ignores one of the time-dimensions. The solution is to use "diagonal" truth-conditions, on which, for example, the ordinary future tense tracks what is happening later in "both time dimensions simultaneously."

There are a couple of ways to write down truth-conditions that do this. The most straightforward way yields a truth condition like this:

(3)   "The square will be white" super-is true at T if and only if it super-will be the case that (the square super-is white at a time later than T).

To have a label for supertensed truth-conditions like these let's call them weak supertensed truth-conditions.

There is something weird about these truth-conditions. Consider the scenario depicted in Figure 4.8. Now this scenario is itself weird, and in fact is incompatible with MST-Supertense as we have been understanding it. As we have been understanding it, the only thing that changes as time passes is which time super-is present. But in Figure 4.8 there is also a change in what color the square is in 2012 (and later times). It super-is black in 2012, but super-will be white in 2012.

But let's look past the weirdness of the scenario. The scenario certainly makes sense. Notice that if (3) is correct then, in the scenario, "The square will be black" super-is true in 2012. So if (3) is correct then the fact that the square super-is white, not black, at times later than 2012, is completely invisible to ordinary tensed language. This is true in general. The truth-conditions for ordinary tensed language we are contemplating leave ordinary tensed language unable to talk about a huge chunk of reality.

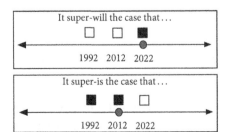

Figure 4.8 Further adventures of a square.

Once noticed, this should not be surprising. After all, we are trying to map our one-dimensional ordinary tensed talk onto the two time-dimensions in the moving spotlight theory. Most of what is going on in those two dimensions is going to be left out.

This is bad; but how bad? Its badness is mitigated by the fact that, according to the theory, the scenario in Figure 4.8 is impossible. And in Figure 4.7 the color the square super-is in 2012 is the same as the color it super-will be in 2012. So it is not like there are changes going on that ordinary tensed talk is blind to.

I said that there are two ways to give diagonal truth-conditions to ordinary tensed talk in MST-Supertense. According to the first way, the weak truth-conditions, people standardly use, for example, the ordinary present tense to talk about the time at which they are speaking. "I am hungry" super-is true at a time T if and only if I super-am hungry at T. Some believers in the moving spotlight might prefer an approach that says that the ordinary present tense is a way to talk, not about the time at which one is speaking, but about the time that is present. A second kind of truth-condition for "I am hungry" does this:

"I am hungry" super-is true if and only if I super-am hungry at the time that is present.

These are the strong truth-conditions for ordinary tensed language. Note that MST-Supertense with strong truth-conditions denies what MST-Supertense with weak truth-conditions accepted, namely that ordinary tensed sentences, like "I am hungry" and "The square is white," super-are true at some times and false at others. According to the strong truth-conditions they super-are just plain true, or just plain false. To make the comparison easier, here is the strong truth-condition for the other example sentence I have been using:

"The square will be white" super-is true if and only if it super-will be the case that (the square super-is white at the time that super-is present).

The strong truth-conditions require more to happen for "The square will be white" to be true than the weak ones do. On the weak truth-conditions, "The square will be white" super-is true whether or not "presentness moves." Even if presentness super-will always be stuck at one time, "The square will be white" can be true. On the strong truth-conditions presentness needs to move to a time at which the square is white for this sentence to be true. This difference speaks in favor of the strong truth-conditions. Time must pass for any tensed sentence to be true.

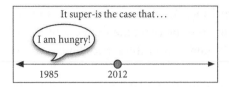

Figure 4.9 A problem for the strong truth-conditions?

However, the theory says that time does pass. Presentness does move into the future. So, given the rest of the theory, the strong and weak truth-conditions are equivalent.

How else might one go about trying to decide between the weak and the strong truth-conditions? Some might think the following kind of argument favors the weak truth-conditions. Suppose I super-say "I am hungry" in 1985, and in fact I super-am hungry in 1985 (see Figure 4.9). Despite this the sentence I said super-is (just plain) false if we use the strong truth-conditions, since it is 2012 that super-is present. That might seem bad.

But I do not think this is a big problem for the strong truth-conditions. If they are correct then, although the sentence I said super-is false, it super-was true, and moreover it super-was true "when the time at which I said it was present" (that is, it super-was true that (1985 super-is present and I super-am hungry in 1985)).

Nothing in what follows is going to turn on whether we put the weak or the strong truth-conditions into MST-Supertense, so I will not discuss arguments for one or the other of them any further.

MST-Supertense avoids talking about supertime by using tense operators. But, as we will see, those tense operators raise some problems of their own, in particular in connection with what the theory says about the rate at which time passes. So I want to present a final version of the moving spotlight theory, one that does without supertime and also does without primitive tense operators.

4.3 The Moving Spotlight via Temporal Perspectives

In all moving spotlight theories there is change in which time is present. Different theories analyze this claim in different ways. In MST-Supertime supertime plays a key role in the analysis. That theory, again, says that for there to be change in which time is present is for different times to be present relative to different points in supertime. But why do we need supertime to play this role? Why not let time play it? A theory that does will say

(4) Each time is present relative to itself, and only to itself.

In this section I will develop a theory that says (4), and call it the moving spotlight theory with time-as-supertime (MST-Time).[14]

Already, just by saying (4), MST-Time faces a challenge. Is it really a version of the moving spotlight theory in the first place? "Actually," one might object, "it is just the block universe theory in disguise. After all, the block universe theory says that each time is simultaneous with itself. And isn't that the same as saying that each time is present relative to itself?"[15]

No, it is not. The claim

> For it to be the case that T is present relative to S is for it to be the case that T is simultaneous with S.

is part of reductionism about A-vocabulary. All versions of the moving spotlight theory reject this thesis.

The best way to get a firmer grip on how MST-Time differs from the block universe theory is to switch one's attention from sentences like "T is present relative to T" to others. Now what MST-Supertime takes to be relative to points in supertime MST-time takes to be relative to instants of time. So step back and ask, What does MST-Supertime take to be relative to points in supertime? The only sentences of MST-Supertime we have seen so far that explicitly relativize anything to points in supertime are sentences like "Time T is present at, or relative to, supertime P." This is because in the version of MST-Supertime I have been working with the only thing that varies from one point in supertime to another is which time is present. Still, the language in which the theory is formulated should be understood to leave conceptual room for the idea that other things vary from supertime to supertime. We should be able to say

> Relative to supertime P1, George is hungry at noon on 5/12/46; but relative to supertime P2, George is not hungry at noon on 5/12/46.

---

[14] When I first wrote about MST-Time, in "On the Meaning of the Question 'How Fast Does Time Pass?'," where I called it MST-TST, I worried that it did not really contain objective becoming. In the terminology I have been using in this book the worry was that in MST-Time time does not undergo robust change. But, as I said earlier, I no longer think that we should insist that robust passage requires robust change. In that earlier paper I also described yet another version of the moving spotlight theory (which I called MST-PT2) with fundamental tense operators. I said that that theory stands to MST-Time the way that MST-Supertense stands to MST-Supertime. In the paper I worried that MST-PT2 was not really an intelligible theory. My doubts that it is intelligible have grown, but so have my doubts that it is superior to MST-Time. I will not discuss it in this book.

[15] I do not know the full history of this worry. Something like it appears in Schlesinger, "The Stream of Time" (p. 259)—he does not endorse it—and in Horwich, *Asymmetries in Time* (p. 22), and Zimmerman, "The A-Theory of Time, The B-Theory of Time, and 'Taking Tense Seriously'," (p. 451)—they do.

Of course, sentences like this should come out false in MST-Supertime. I am just saying that even if false they should be consistent.[16] So in MST-Supertime "George is hungry at noon on 5/12/46" is not metaphysically complete. To make it metaphysically complete we need to prefix it with something of the form "Relative to supertime P."

So in MST-Supertime we can associate with any point in supertime P the collection C(P) of sentences S such that $\ulcorner$Relative to X, S$\urcorner$ is true, relative to the assignment of P to "X." The collection C(P) says what reality is like from the "perspective" of supertime P. Each point in supertime is, or "provides," a perspective on reality. Pictures like those in Figure 4.10, which show what time is like relative to various points in supertime, could in principle be expanded, to show what spacetime and all of its contents are like relative to various points in supertime.

In fact we should do more. We should take points in supertime to provide perspectives, not just on other things, but also on themselves. So a complete picture of what reality is like from the perspective of a point in supertime will also illustrate what supertime is like from that perspective. (Of course, since the way supertime is does not vary from one supertime to another there is rarely any need to make this explicit.)

In MST-Time time plays the role supertime plays in MST-Supertime. So "George is hungry at noon on 5/12/46" is not metaphysically complete in MST-Time either. To get a complete statement we need something of the form "Relative to time T, George is hungry at noon on 5/12/46." Similarly, where MST-Supertime says that it is consistent, though false, to say

Relative to supertime P1, George is hungry at noon on 5/12/46; but relative to supertime P2, George is not hungry at noon on 5/12/46,

MST-Time says that it is consistent, though false, to say

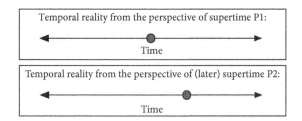

Figure 4.10 How reality is, from various perspectives in supertime.

---

[16] Here I am just repeating a point I have already made, in the context of a different theory. In Section 4.2 I said that it should make sense in MST-Supertense to say that the square super-is black in 2012 but super-will be white in 2012.

(5) Relative to 4pm on 8/8/03, George is hungry at noon on 5/12/46; but relative to 5pm on 8/8/03, George is not hungry at noon on 5/12/46.

Here, then, is something else that distinguishes MST-Time from the block universe theory. In the block universe theory if sentences of the form "Relative to T1, S at T2" make sense at all, they are equivalent to sentences of the form "S at T2." So in that theory (5), if it makes sense at all, is inconsistent.

In MST-Time we can associate with any time T a collection C(T) of sentences S such that ⌜Relative to X, S⌝ is true (and metaphysically complete), relative to the assignment of T to "X." The collection C(T) says what reality is like from the perspective of T. Each time is, or "provides," a perspective on reality. And each perspective is, among other things, a perspective on time itself. From the perspective of each time it and it alone is present. Since different times are present from different perspectives, time itself changes. The pictures we should associate with MST-Time, then, are the pictures in Figure 4.11. That figure associates a different picture of time as a whole with each instant in time. The dot indicates which time is present, and the dashed arrows go from a time to the time(s) simultaneous with it.

Figure 4.11 helps make clearer that, in MST-Time, it is one thing for T to be present relative to T, and quite another for T to be simultaneous with T. For one thing, T is not simultaneous with T full-stop; it is simultaneous with itself relative to each time. For another thing, which time is present changes from perspective to perspective, but which times are simultaneous with which does not.

I have said that all versions of the moving spotlight theory reject reductionism about A-vocabulary. But not all of them reject reductionism about tense, narrowly construed. MST-Supertense does, but MST-Time does not.

MST-Time is strange because there are two things that "at T" can mean in the theory. "At T, . . ." can abbreviate "relative to T," or—perhaps this is a more illuminating locution—"from the perspective of T, . . ." This is what the first temporal modifier means in "At 4pm, 8/8/13, George is hungry at noon on 5/12/46."

Figure 4.11 Pictures of temporal reality in MST-Time.

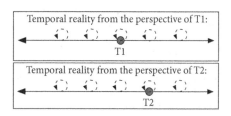

But "at T" can also mean what "at T" means in the block universe theory. That is what the second temporal modifier means in this sentence.[17]

What about tensed sentences? What should MST-Time say is the effect of tensed verbs in our ordinary use of natural language? We faced the same question in MST-Supertense. There I discussed two kinds of truth-conditions the theory could give tensed language. MST-Time can give analogues of those weak and strong truth-conditions for tensed language.

Start with the weak truth-conditions. In the block universe theory an ordinary tensed sentence has its truth-value relative to a single time. In MST-Time the weak truth-conditions say that a (typical) ordinary tensed sentence has its truth-value relative to two times. Here is a sample weak truth-condition:

> "The square will be white" is true at T from the perspective S if and only if there is a time U later than T and a perspective V later than S such that the square is white at U from perspective V.

In MST-Supertense the strong truth-conditions say that an ordinary tensed sentence super-has its truth-condition absolutely, not relative to a time. Here, in MST-Time, the strong truth-conditions say that an ordinary tensed sentence has its truth-condition relative to a single time—as in the block universe theory. But the truth-condition is not the one we find in the block universe theory. Instead, we get this:

> "The square will be white" is true from the perspective of T if and only if there is a perspective U later than T such that, from the perspective of U, the square is white at the time that is present from the perspective of U.

Now I admit that this sentential operator "From the perspective of T . . ." is strange. Let me dwell a little longer on how it works.

I have said that in MST-Time each time provides a perspective on reality. Now even in the block universe theory it makes sense to talk of times providing perspectives on reality. But in that theory the relationship between the perspectives and the reality they are perspectives on is different, and much simpler.

---

[17] What does "at T" mean in the block universe theory? Some believers in the block universe say that "at T" modifies the subject. "George is hungry at T" amounts to "The T-temporal part of George is hungry." Believers in the block universe who reject temporal parts will say something different. Some say that "at T" modifies the predicate "is hungry," others that it modifies the verb "is." (These are hypotheses about "at T" that metaphysicians have discussed. I do not know what linguists who presuppose the block universe theory say. Doubtless it is more sophisticated (and closer to the truth). I will have more to say about the relationship between the block universe theory and the doctrine of temporal parts in Chapter 10.)

Figure 4.12 Two viewers of a cylinder.

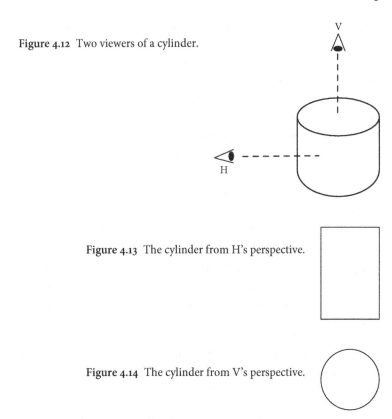

Figure 4.13 The cylinder from H's perspective.

Figure 4.14 The cylinder from V's perspective.

Here is an analogy with a more literal use of "perspective"—visual perspective. Suppose two observers, H and V, are looking at a cylinder (see Figure 4.12). H is looking at it from the side-on, while V is looking at it from above.

The cylinder has a certain shape from H's perspective and has a different shape from V's perspective. From H's perspective (Figure 4.13) it is rectangular while from V's perspective (Figure 4.14) it is circular. (Suppose that H and V each have only one eye and that all other cues for judging depth are absent.)

In this case the way "From H's perspective. . ." works is straightforward. Facts about how things are from H's perspective are determined by (i) the "perspective-independent" facts about the cylinder's shape, and (ii) the perspective-independent facts about how H is positioned in space relative to the cylinder.

If a believer in the block universe wanted to use operators like "From the perspective of T . . ." in his theory the most natural way to do so is to say the analogous thing. Whether it is the case that from the perspective of T,

such-and-such is true, is determined by the perspective-independent facts together with facts about T's location (temporal location) in reality.[18]

What makes MST-Time different from the block universe, and what makes it
such an interesting theory, in my view, is that facts about what is true from the
perspective of some time T are not determined by the perspective-independent
facts. In fact, not only does MST-Time deny that perspectival facts are determined by the perspective-independent facts; the theory also says that there are
no perspective-independent facts. While it is true that from the perspective of T,
T is present, this does not follow from any perspective-independent fact about
time together with some facts about how T is related to other times.

Call this feature of MST-Time "perspectivalism." I have stated perspectivalism
as the view that there are no perspective-independent facts. But this way of stating
the view can be misleading. Isn't the fact that T is present from the perspective of
T a perspective-independent fact? Well, maybe. It might depend on how exactly
one wants to use fact-talk. Since I think that different ways of using fact-talk are
equally good, it would be good to have a characterization of perspectivalism that
does not use fact-talk. Here is a first-pass:

(6) The only metaphysically complete sentences are sentences that have as
their main connective something of the form "From the perspective of
T . . ."

This, however, cannot quite be right. "From the perspective of 1950, 1950 is
present, and from the perspective of 2000, 2000 is present" should count as metaphysically complete, but it is not of the right form. Fixing this problem is easy.
Just have separate conditions for a sentence to be metaphysically complete, and
a sentence to be a "basic" metaphysically complete sentence:

(7) The only basic metaphysically complete sentences are sentences that have
as their main connective something of the form "From the perspective of
T . . ." A sentence is metaphysically complete if and only if it is either basic,

---

[18] If we want to use temporal perspectives talk in the block universe theory, how precisely should
that talk be understood? That depends on how we want to use perspective talk. We could use it
the same way it is used in MST-Time, and go in for saying things like "From the perspective of T1,
X is F at T2." Then we say that the "From the perspective . . ." operator is redundant; sentences
starting with it are true just in case the sentences the operator attaches to are true. Alternatively, the
block universe theory could use perspective-talk differently from the way it is used in MST-Time. It
could say that only if S is a tensed sentence does ⌜From the perspective of T, S⌝ make sense. Then it
should say that, for any tensed sentence S, ⌜From the perspective of T, S⌝ is true just in case S is true
at T. (We have already seen how S's truth-value at T is determined by the perspective-independent
facts (which in this theory are the tenseless facts) together with facts about T's temporal location in
reality.)

or built up from basic sentences using sentential connectives ("if," "and," "or," "not," and so on).

But (7) is still not right. I want MST-Time to say that I exist from the perspective of 2000 and from the perspective of 2050. And I want this to entail that something exists from the perspective of 2000 and the perspective of 2050. And "Something exists from the perspective of 2000 and of 2050" should be metaphysically complete. But it does not qualify as such under (7). The solution is simple. Revise perspectivalism again to say

(8)   The only basic metaphysically complete sentences are sentences that have as their main connective something of the form "From the perspective of T . . ." Now say that a basic metaphysically complete open sentence is the result of replacing one or more occurrences of a name in a basic complete sentence with (possibly distinct) variables. Then a (closed) sentence is metaphysically complete if and only if it is either basic, or built up from (closed or open) basic sentences using sentential connectives and quantifiers.

(Note that (8) also counts as basic sentences like "For every T, from the perspective of T the year 2000 is after the year 1950"—that is, 2000 is after 1950 from every perspective.)

Statement (8) is a slightly softened-up version of perspectivalism. It allows that, so to speak, there are perspective-independent facts about existence. We are not restricted to saying merely that from each perspective, some things exist. If we were restricted to saying this we could not go on to ask, do the same things exist from the perspective of S and from T? Without the restriction this has an answer. It could be that there are some things such that, from each perspective, they exist.

I think there are good reasons to use (8) rather than (7) when stating MST-Time, but it is true that once we move from (7) to (8) lots of questions about the relationship between perspectival and perspective-independent facts press themselves upon us. Given that there are perspective-independent facts about existence, are there also perspective independent facts about identity and distinctness? Should we be able to say, for example, that there are at least two things that exist from many perspectives? It certainly seems we should. So (8) should be further revised to allow statements of the form "X = Y" and "X ≠ Y" to be metaphysically complete.

But this is just the beginning. Here are two more natural questions. If, from the perspective of T, something is F, does it follow that there is something such that, from the perspective of T, it is F? I think it should. (This is the analogue, for perspectives, of the modal Barcan formula: if it is possible that there be an F, then there is something that is possibly F.) But I do not think that MST-Time should say that if something exists from some perspective then it exists from every perspective.[19]

The second question is, are facts about logical and mathematical truths perspective-independent? This question illustrates the fact that "perspective-independent fact" is a tricky phrase (another reason I prefer to avoid it). To be consistent with my explanation of "there are no perspective-independent facts" I should interpret this question to mean

(Q1) Are sentences expressing truths of logic and mathematics, like "2+2 = 4," metaphysically complete?

This interpretation equates perspective-independence with truth outside a perspective. This is strong perspective-independence. There is a weaker kind of perspective-independence. A truth is weakly perspective-independent if and only if it is true from every perspective.[20] Interpreted to be about this notion the question becomes

(Q2) Are sentences expressing truths of logic and mathematics, like "2+2 = 4," true from every perspective?

I certainly want to answer yes to (Q2). Maybe there is some pressure to answer yes to (Q1) once one has answered yes to (Q2). Maybe we should say that anything that is weakly perspective-independent is also strongly perspective-independent. That would lead to a much less radically perspectival version of MST-Time. There would be tons of strongly perspective-independent facts. But facts about which time is present would still be perspectival, and these perspectival facts would still not be determined by the perspective-independent ones.

I think that believers in robust passage are best off denying that everything that is weakly perspective-independent is strongly perspective-independent. I think that accepting this makes the few remaining perspective-dependent facts feel a little unreal, a little ghostly. And as this happens the passage of time in the theory becomes less robust.

---

[19] So the theory should not contain all instances of the perspectivalist version of the converse Barcan formula: if there is something that, from some perspective, is F, then from some perspective, there is something that is F. (Specifically, the instance in which "is F" is replaced by "does not exist.") The fact that, in this theory, there may be variation in what exists from perspective to perspective will come up again in Section 6.5.

[20] The distinction between weak and strong perspective-independence is analogous to Kit Fine's distinction between two kinds of necessity: truth whatever the circumstances and truth regardless of the circumstances. See *Modality and Tense* (p. 324).

When I introduced MST-Supertime I discussed the objection that MST-Supertime does not contain objective becoming because in it time does not undergo robust change. The same objection could be made about MST-Time.[21] In neither theory are sentences of the form "T is present" metaphysically complete. Either they are true only relative to points in supertime or true only from the perspective of instants of time. My reply to this objection is the same as before. We should not automatically reject a theory of objective becoming because it fails to contain robust change. MST-Time is a fantastic departure from the block universe. The passage of time in it is more substantial than the anemic passage that goes on in the block.

(Sometimes the objection is put like this: for objective becoming to occur presentness must be a monadic, or an intrinsic, property. I think this is actually a different objection. I think that presentness can be intrinsic—and, in MST-Time, is intrinsic—even though "T is present" is not metaphysically complete. Intrinsicness and time-relativity are not incompatible.[22] So I do not think this objection applies to MST-Time.)

Other philosophers have proposed versions of the moving spotlight theory that resemble MST-Time. The theory it most closely resembles is Kit Fine's first form of "non-standard" realism about tense (see his "Tense and Reality"). He describes the view like this:

reality at another time is an *alternative* reality. It is neither a facet of the one true reality nor a hypothetical determination of the one true reality, but another reality on an equal footing with the current reality; and the facts belonging to such a reality are full-fledged facts, sharing neither in the incomplete status of a facet nor in the insubstantial character of a hypothetical fact. (P. 279.)

Later Fine writes that in non-standard realism

there is no saying how reality is without presupposing a temporal standpoint from which the description is given. (P. 280.)

These remarks make clear that Fine's theory and MST-Time are animated by the same spirit. I am, however, not sure whether they are the same theory.[23]

[21] Dozens of philosophers have said something like this about a theory like MST-Time. Among them are Horwich, *Asymmetries in Time* (p.22); Callender, "Shedding Light on Time" (p. S591); Zimmerman, "The Privileged Present" (p. 212); and a former version of myself, in "Relativity and the Moving Spotlight."

[22] See Haslanger, "Endurance and Temporary Intrinsics" for a defense of this view. (Spencer defends the idea that monadicity and time-relativity are compatible in *Relativity and Ontology*.)

[23] In a footnote on page 279 Fine mentions some other philosophers, most importantly Dummett, in "A Defense of McTaggart's Proof of the Unreality of Time," (p. 503), and Horwich, in *Asymmetries in Time* (pp. 23–4), who also discuss views that are similar to (but, I think, distinct from) MST-Time.

I characterized perspectivalism, in (8), using the notion of metaphysical completeness. Some passages in Fine's paper may suggest that that characterization does not adequately capture the kind of perspectivalism I want MST-Time to embody. But it is more important to me that MST-Time be understood as a kind of perspectivalism that marks a radical departure from the block universe theory, than it is that my way of understanding this kind of perspectivalism is correct. Everything I say about MST-Time in what follows should be compatible with alternative analyses of its form of perspectivalism.[24]

I think the best way to get the hang of how MST-Time works and how it differs from the block universe is to use the theory. That will happen in Chapters 6 and 7, when I look at some famous arguments against the moving spotlight theory.

4.4 Appendix: MST-Time with Absolute Presentness

It looks like we face a trade-off. If we want there to be "absolute" facts about which time is present, if we want sentences like "Noon, New Years' Day, 2013 is present" to be metaphysically complete, then we should go for MST-Supertense. If we dislike supertense, though, and want a version of the moving spotlight theory that does without it, then we should go for MST-Time. But that theory lacks absolute facts about which time is present.

There is a version of MST-Time that does give us both things. Take the version of the theory I just described, with its perspective-dependent facts about which time is present. Then add to it a (strongly) perspective-independent fact about which time is present. This version says not just that for each time T, from the perspective of T, T is present. It also says, flat-out, that there is exactly one time that is present, absolutely speaking. (Presumably the theory should also say, of each thing that exists, what that thing is like, absolutely speaking, outside of all perspectives.)

I will mostly ignore this version of MST-Time, though it will come up a few times in later chapters.

4.5 Appendix: MST-Time and Semantic Relativism

I am trying to avoid proposition-talk as much as possible, but some recent discussions of whether propositional truth is relative or absolute raise again the

These are views that, in one way or another, reject the idea that there are perspective-independent facts. Fine expresses reservations about whether these views are the same as the one he is proposing.

[24] See Solomyak, "Actualism and the amodal perspective," for a different attempt to articulate what "pluralism about perspectives" might be. Thanks here to Ted Sider.

worry that MST-Time is just a version of the block universe theory. As I mentioned in the second appendix to Chapter 2, some believers in the block universe say that propositions have their truth-values only relative to times. So what, if any, is the difference between a version of the block universe theory like this, and MST-Time?

Looking more closely at one particular relativist makes this question more pressing. John MacFarlane's basic way of using truth-talk, in his discussion of the semantics of tense in "Truth in the Garden of Forking Paths," is to say things of the form "sentence S, as used at T, is true as assessed from U." He links this "assessment-sensitivity" of sentences to relative truth for propositions. Sentence S, as used at T, is true as assessed from U is true if and only if S, as used at T, expresses a proposition that is true relative to U. His way of speaking about sentence truth looks a lot like what happens in MST-Time, which affirms things like "George is hungry at T from the perspective of U." (Strictly speaking MacFarlane relativizes truth to circumstances of evaluations, not times; but each circumstance determines a time.)

MST-Time is not just a notational variant of MacFarlane's theory. As far as I can tell, MacFarlane thinks that a tenseless description of reality has a truth-value that does not vary with time. This is false in MST-Time.

MST-Time need not incorporate semantic relativism at all. True, the weak truth-conditions for tensed sentences in MST-Time relativize truth to pairs of times. But they say nothing about propositions. According to the weak truth-conditions, "George will be hungry" is true at T from perspective U if and only if George is hungry at a time later than T from a perspective later than U. MST-Time could go on to say that a token of "George will be hungry" that exists at T expresses one proposition from perspective U, a proposition that is just plain false, and a different proposition from perspective V, a proposition that is just plain true.

It may be that the best version of MST-Time is a relativist one. My point is that if it is, it is a far more radical version of relativism than the relativist treatments of tensed language that have been proposed. What makes it radical is that its way of being relativist requires a departure from the block universe theory.

5

Growing Blocks and Branching Times

5.1 Two More Theories of Robust Passage

The moving spotlight theory says that time changes. This change constitutes the robust passage of time. I am going to spend the rest of this book discussing this theory, and the block universe theory. But the moving spotlight theory is not the only theory of time in which time changes. There are two others that are often discussed: the "growing block" theory of time, and the "branching time" theory of time.

There is, to my mind, only one reason to even be tempted to prefer these theories to the moving spotlight theory. The "reason" is that they better capture the idea that the future is open. But in fact they do not capture this idea better.

First, the theories. It is easiest to present the growing block theory in the same framework I used to present MST-Supertense. So the language in which I will express the theory will contain supertensed verbs. The growing block theory is also easiest to explain if we take the 4D view, rather than the 3 + 1 view, of spatiotemporal reality, and speak in terms of spacetime, and of instants of time as parts of spacetime.

The theory says that spacetime extends infinitely into the past, but only finitely far into the future. Spacetime is "truncated." And there is one instant of time, the one on the "future edge" of spacetime, that is later than all the others. That time is the present.

The theory also says that spacetime changes. The time that super-is present will not super-always be present. The truncated block universe super-will grow in the future direction. More precisely, the theory says

There super-is an instant of time M later than all other instants. But while all instants super-will continue to exist, and be in the same order (if T1 super-is earlier than T2, then it super-will continue to be earlier than T2),

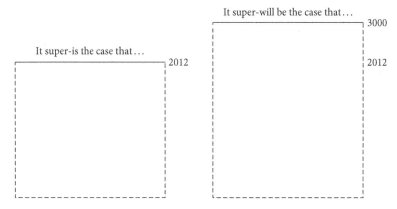

Figure 5.1 The growing block.

there super-will also be some "new" instants, one of them later than all other instants, new and old (including M).[1]

This change in the size of spacetime constitutes the robust passage of time.[2] The picture that goes with the growing block theory is in Figure 5.1.

Now for the branching time theory. Thinking in terms of spacetime makes the branching time theory difficult to picture, so let us retreat again to the 3 + 1 view, ignore space, and focus just on time. The branching time theory says that time extends into the past like a line, but extends into the future like a "garden of forking paths" (see Figure 5.2). While there is only one instant corresponding to noon, New Year's Day, in the year 800, there are many instants, one for each "branch" or "path," corresponding to noon, New Year's Day, in the year 3000. One time is on the "boundary" where time starts to fork. That time is the present.[3]

The theory also says that time changes. The time that super-is present super-will not always be present. All but one of the branches near the present

---

[1] This is not a complete statement of the theory, but it is all we will need. (A complete statement will say more about how the "new" parts of spacetime are connected to the old part, and also say "how long it super-will be" until there is a time that is both later than all other times and is one second later than M.)

[2] C. D. Broad invented the growing block theory of time; see *Scientific Thought* (p. 66). It has several contemporary defenders, including, for example, Michael Tooley, in his book *Time, Tense, and Causation*. My formulation of the theory, however, does not follow either of their models.

[3] In Figure 5.2 there are stretches in the future during which time does not branch. But that is an artifact of the diagram; really in the future time is "continuously" branching.

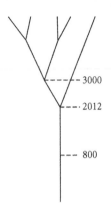

Figure 5.2 A picture of branching time.

It super-is the case that... It super-will be the case that...

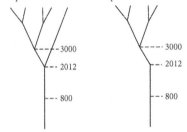

Figure 5.3 Passage in branching time.

super-will "fall off" (go out of existence), leaving a later time on the boundary, as in Figure 5.3. This change in what branches exist constitutes the robust passage of time.[4]

5.2 The Open Future

Here, then, are two more theories with robust passage of time. Why might they seem better than the moving spotlight theory? Because, it might be thought, they

---

[4] Storrs McCall is a contemporary defender of the branching time theory; see for example his paper "Objective Time Flow."

There are theories that speak of branching time that are distinct from the one that is my target. One is a version of the block universe theory that says that spacetime has a branching structure. It does not say that branches continually go out of existence. Since this is a version of the block universe theory it is not a theory of robust passage.

In other theories the thing that has a branching structure is some abstract surrogate for time, not time itself. Time itself does not branch. The abstract surrogate is just there to "model" tensed language, especially language about future-indeterminacy. I am not sure how or whether this view differs from the version of MST-Supertense I will describe shortly. (Pooley discusses a view like this in "Relativity, the Open Future, and the Passage of Time.")

contain an important feature of the (robust) passage of time that the moving spotlight theory lacks. Here is one way to put the thought:

The future is "open," "mere potentiality," while the past is "closed" or "determinate." And a stretch of time changes from being "open" to being "closed" when it changes from being future to being present. But in the moving spotlight theory the future and the past are both closed. The transition from being open to being closed, from being mere potentiality to being actuality, is missing.

Now one might doubt whether there is any connection between the passage of time and the open future. But let us suppose there is and see what follows. Why think that the future is "closed" in the moving spotlight theory and "open" in the other two theories? A natural answer goes like this (it may help to consult Figure 5.4 while you read):

In the moving spotlight theory the universe has a single, unique future. It is right there: the region of spacetime later than the present is the future, and all sorts of material things (condominiums on Mars, maybe) are located in it. But in the other theories the universe lacks a single, unique future. In the growing block theory the universe has no future at all. That is, there is no such place in spacetime as the future. Spacetime gives out right where the future would start. In the branching time theory, on the other hand, the universe has many futures, one for each path through the garden. To describe what happens in the future in this theory it is not enough to specify where things are at each time. You need to specify where things are at each time in each branch. Maybe in the year 3000 there are condos on Mars in some branches but not in others.[5]

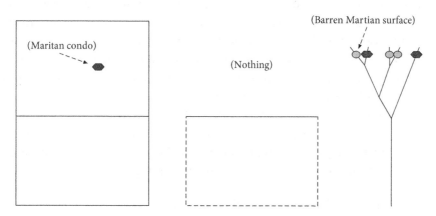

Figure 5.4 The future in the three theories.

[5] There is another difference between the moving spotlight theory, on the one hand, and the growing block and branching time theories, on the other. It is a consequence of the different ways they treat the future. In the moving spotlight theory "present" is a basic piece of vocabulary. But in these other theories "present" need not be basic. It can be defined in other terms, either as "the

Suppose everything in this speech is correct. How does it bear on whether the future is open or closed in each of these theories of time? What is the connection between the number of future "branches" (space)time has and the future's being open? The implicit answer, I think, is this:

To say that the future is "open" is to say that it is unsettled, or indeterminate, what will happen. But if there is such a place in spacetime as the future, and there is exactly one such place, then there must ("already") be a determinate fact of the matter about what will happen. What will happen is what is going on in that place. So the future is not open in the moving spotlight theory.

In the growing block theory, though, the future is open. That is because there is no such place as the future. Instead there is nothing, not even an empty spacetime. Because there is no such place as the future there is nothing there to "settle" what will happen, or what will be the case. There is room for "potentiality to transition into actuality." When a stretch of time (which is just a chunk of spacetime) is added to spacetime, its coming into existence "settles" what happens during that stretch of time. What happens is just whatever is going on in that chunk of spacetime.

And in the branching time theory the future is open. That is because there are many future regions (branches) of spacetime. The fact that different things happen in different futures means that it is not settled what will happen. When a bunch of branches go out of existence, say all but one of the branches that "begin" before 2020, their going out of existence settles what happens up until 2020. What happens is just whatever is going on in the single remaining branch.

I think this is all wrong.

It is just not true that there is no room for indeterminacy about the future, for "future potentiality" to become "actuality," in the moving spotlight theory. True, the way I have been talking about the theory leaves no room for this. I have been talking as if there either determinately are Martian condos located in the year-3000 region of spacetime or there determinately are not. (This is what the block universe theory says, and I have been discussing the version of the moving spotlight theory that is the "minimal departure" from the block universe.) But the moving spotlight theory can easily be modified so that the future is indeterminate. Let us see how to modify MST-Supertense. We just need to have the theory affirm the following about this example:

(1) It super-is indeterminate whether there super-are Martian condos in the year 3000.

instant later than all others" or "the last instant before time begins to branch." (These theories, then, may affirm reductionism about A-vocabulary. Since they use supertensed verbs they must still reject reductionism about tense narrowly construed.) But I don't think this fact makes these theories better than the moving spotlight theory.

(2) It super-is (also) indeterminate whether there super-will be Martian condos in the year 3000.

The theory should also then say that what super-is indeterminate super-will
become determinate:

(3) It super-will be the case that (it super-is determinate whether there are
Martian condos in 3000).

Claim (1) is needed in addition to (2) to ensure consistency with the principle
that "what super-is determinate always super-will be determinate." Figure 5.5
illustrates what is going on.

In light of (1) through (3) is it not difficult to see what more general theses
about time and indeterminacy the theory should contain. It should say that when
T super-is in the future, it super-is indeterminate what super-is happening at

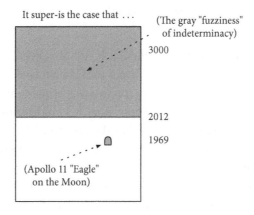

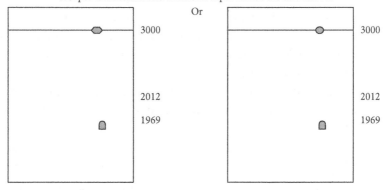

Figure 5.5 The open future in the moving spotlight theory.

T; and that it super-becomes determinate what super-is happening at T when T super-becomes present.[6]

Now one might respond that the branching time theory and the growing block theory are still better theories of the open future. What might make them better? Well, the somewhat obscure claim that "the future is open" has been explained using the concept of indeterminacy. Ask yourself, what is indeterminacy? Set the growing block theory aside for now and think about the branching time theory. A fan of that theory might say

My theory has a natural answer to the question, What is indeterminacy? I can give a "reductive analysis" of indeterminacy. That is, with my theory I can give truth-conditions for sentences containing "indeterminate" in a more basic language that does not contain "indeterminate." For example,

- "It is indeterminate whether there will be condominiums on Mars" is true if and only if there are condominiums on Mars in some but not all future branches.

The moving spotlight theory, on the other hand, cannot say in other, more basic terms what indeterminacy is.

This is not very compelling. First, it is not obvious that all indeterminacy must concern the future. Maybe, if there were some sand on my desk right now, it could be indeterminate whether there was a heap of sand on my desk right now. Maybe quantum mechanics shows that it is indeterminate where anything is. Maybe it is indeterminate whether the sentence "This sentence is false" is true or false. Many philosophers have seen good reasons to think that one or another of these things is indeterminate. But these examples of indeterminacy cannot be understood by reference to what is going on in the many future branches of time. The branching time theory's reductive analysis is not general.

But suppose we ignore these other kinds of indeterminacy. It is also not true that the moving spotlight theory cannot give a reductive analysis of indeterminacy. There are reductive theories out there that are perfectly compatible with the theory. For example, there is the theory that indeterminacy is ignorance. To say that it is indeterminate whether P is to say that it is unknowable (for certain special kinds of reasons) whether P.[7]

However, I grant that the theory that indeterminacy is ignorance is not very popular, and is especially implausible when used to make sense of the claim that

---

[6] In fact, the modified theory could eliminate "present" as a piece of basic vocabulary, and instead define "T super-is present" as "T super-is the latest time with this property: it super-is determinate what super-is happening at it."

[7] Timmothy Williamson is probably the most prominent defender of this theory of indeterminacy. See his book *Vagueness*.

the future is open. The indeterminacy of the future is supposed to be "metaphysical indeterminacy," indeterminacy "in the world." This kind of indeterminacy, it seems, cannot be analyzed in terms of ignorance. So let us turn back to a prior question. Why is it bad for a theory to lack a reductive analysis of indeterminacy? The speech I imagined in defense of the branching time theory just assumed that this is bad. But in fact I think the opposite is the case. I think that indeterminacy has no reductive analysis.

Even if I am wrong and there does turn out to be a correct reductive analysis of indeterminacy, the analysis in the branching time theory is certainly not right.[8] Look back at the branching time in Figure 5.4. On some branches there are condos on Mars in 3000. On others there are not. In light of this "It is indeterminate whether there will be condos on Mars" does not seem like the right thing to say at all. The right thing to say is that there will determinately be condos on Mars. There they are, right there on the "right-most" branch.

To say that the future is indeterminate is to say that the future is "unsettled." But there being Martian condos here in the future, but not there in the future, is not a way for the future to be unsettled. Time may have more than one branch in the future. But those extra branches are just more "determinate matters of fact." And an overabundance of determinateness does not make for indeterminacy.

This complaint is an analogue of a common complaint about Modal Realism.[9] Modal Realism is a reductive theory of possibility and necessity. It says, roughly speaking, that

"It is possible that P" is true if and only if there is a spacetime disconnected from the one we inhabit in which P.

But, the objection goes, what is going on in spacetimes disconnected from our own is "just more actuality." There are no talking donkeys in our spacetime. Suppose that there are none in any spacetimes disconnected from our own either. This complete absence of talking donkeys from reality as a whole would not entail that talking donkeys are impossible. The claim that talking donkeys are possible just is not a claim about what is going on in "far away," or unreachable, places.

Modal Realism gets the subject matter of possibility-talk wrong. My complaint about the branching time theory is that it gets the subject matter of future-indeterminacy talk wrong.

---

[8] Barnes and Cameron anticipate the two arguments to follow in their paper "Back to the Open Future." (They also defend the idea that the open future is to be understood in terms of indeterminacy, and present other arguments against the branching time theory that are independent of its treatment of the open future.)

[9] David Lewis's *On the Plurality of Worlds* is, of course, the foremost defense of Modal Realism.

One might reply that "There will be condos on Mars" just means "There are condos on Mars in the future"; but in the branching time theory there is no such thing as "the" future; there are many futures, one for each branch. Then, the reply continues, the right thing to do is count "There will be condos on Mars" as true if there are condos in every future, false if there are condos in none, and indeterminate otherwise. This reply makes three mistakes. First, I disagree that in the branching time theory there are many futures. There is just one. It just has a different shape than in the block universe theory, on account of the branches. Second, even assuming that it is right to say that there are many futures, on this analysis, "It is indeterminate whether there will be condos on Mars" looks like an example of semantic indeterminacy. The truth of this sentence turns on the fact that "the future" does not have a determinate denotation. But the indeterminacy of the future is supposed to be metaphysical, not merely semantic, indeterminacy.

The third mistake in this reply concerns its treatment of tense. The claim that "There will be condos on Mars" just means "There are condos on Mars in the future" must really be the claim that it means "There are condos on Mars in the future." That might be plausible to someone who accepts reductionism about tense. But the branching time theory (and the growing block theory too) reject reductionism about tense.

This third mistake is a big one. It was already there in the earlier proposed reductive analysis of future indeterminacy, which did not have the middle-man about there being condos in "the" future:

- "It is indeterminate whether there will be condominiums on Mars" is true if and only if there are condominiums on Mars in some but not all future branches.

This makes most sense as an attempt to give a tenseless truth-condition to a tensed sentence:

(6) "It is indeterminate whether there will be condominiums on Mars" is true at T if and only if there are condominiums on Mars later than T in some but not all branches.

But tensed sentences do not have tenseless truth-conditions in the branching time theory.

Is there any way to modify (6) so that it is compatible with the branching time theory? The sentence-form we are trying to analyze is "It is indeterminate whether (it will be the case that (X))." The first choice-point is whether the "will" in this form is the ordinary tensed form of "to be" or the supertensed form. Suppose it is the ordinary form. Then one thought is to modify (6) like this:

(7) "It is indeterminate whether there will be condominiums on Mars" super-is true at T if and only if there super-are condominiums on Mars later than T in some but not all future branches.

I don't think this is plausible at all. Suppose things super-are, and super-will be, as in Figure 5.6. There super-are Martian condos on one (the "right") branch and not on the other in the year 3000. But there super-will be only one branch in 3000, and it will have Martian condos on it. (And no times other than 3000 are relevant.) By (7), "It is indeterminate whether there will be condominiums on Mars" super-is true at times before the branching. But that hardly seems like the right thing to say, in light of the fact that it is also true that there super-will be condos on Mars on every branch (in the only relevant time, the year 3000).

What the theory needs to say, I think, is this. Not only super-is (8), a sentence with an ordinary tensed verb, true at times before the branching:

(8) It is indeterminate whether there will be condominiums on Mars.

In addition, (9), a sentence with supertensed verbs, is (just plain) true:

(9) It super-is indeterminate whether there super-will be condominiums on Mars in the year 3000.[10]

Figure 5.6 A branching time scenario.

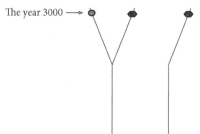

It super-is   It super-will be
the case that... the case that...

The year 3000 ⟶

---

[10] What about the sentence I used earlier, "There super-will be only one branch in 3000, and it will have Martian condos on it"? Should the theory say that while (9) is true this sentence is false? If so then it should also say that "It is not the case that there super-will be only one branch in 3000, and it will have Martian condos on it" is false. Then we are heading toward a theory of indeterminacy that abandons classical logic.

What to say about the relationship between ⌜It is indeterminate whether P,⌝ P, ⌜not-P,⌝ and ⌜P or not-P⌝ is a big topic, too big to weigh in on here, even if I had a firm opinion. My point is just that the branching time theory needs to use the word "indeterminate" and its cognates in its fundamental, supertensed language. It cannot confine the use of that term to a less fundamental language containing only ordinary tensed verbs.

But the use of "indeterminate" in (9) cannot be analyzed as "truth on only some future branches." For what would the analysis look like? It would have to say something like:

- "It super-is indeterminate whether there super-will be condominiums on Mars in the year 3000" super-is true if and only if there super-are Martian condos in the only year-3000-branch, on some but not all year-3000-branches.

This is either nonsense or false, but (9) is supposed to be true.

The conclusion is that the branching time theory needs a notion of indeterminacy that cannot be reductively analyzed as truth on some but not all branches. But then the theory does not have a reductive analysis of all claims about future indeterminacy. It thus has no advantage over the moving spotlight theory in this respect.

I have been arguing that the branching time theory does not really have any advantages over the moving spotlight theory. I have not said anything about the growing block theory. Is it any better?

The growing block theory can be seen as an over-reaction to the problems the branching time theory faces. If it seems right that adding extra future branches that lack Martian condos does not make it indeterminate whether there will be Martian condos, it might seem that the thing to do is to get rid of the future altogether. This might suggest a reductive analysis of future determinacy (and thereby of future indeterminacy) that says, for example,

- "It is determinate that there will be condominiums on Mars" is true at T if and only if there is a time later than T in which there are condominiums on Mars.

Since there are no future times, both "It is determinate that there will be Martian condos" and "It is determinate that there will not be Martian condos" are false at the present time, which is enough for "It is indeterminate whether there will be Martian condos" to be true at the present time.

But the growing block theory has the same fundamental problem that the branching time theory has. It will need to say that while there are no future times, there super-will be.[11] And if the theory is to say that it is indeterminate what those future times super-will be like, it will need to use a notion of indeterminacy that has no reductive analysis.

---

[11] That is, the theory will need to say that the time later than all other times is such that there super-will be times later than it. (The theory certainly should not endorse the other reading of "there super-will be future times"—on that reading it says that it super-will be the case that [there are times later than the time later than all other times], which is a contradiction.)

5.3 Appendix: A Remark on the Structure of Theories of Time

Theories of time make two kinds of claims. They make claims about what exists (ontological claims), and they make claims about how tensed language is to be analyzed. The block universe theory says that there exist past and future times, and also says that tensed language can be given tenseless truth-conditions. MST-Supertense agrees with the block universe theory's ontology but rejects its analysis of tense. Presentism and the growing block theory reject both.

It is important to keep both parts of a theory in mind when evaluating it. If you focus on a theory's ontological claims while foisting on it a foreign analysis of tense the theory will look easy to refute and you will expect those who accept it to behave very differently than they in fact do. This is what is going on when David Lewis says that no one really believes the growing block theory:[12]

Consider the philosophers who say that the future is unreal. It is hard to believe they mean it. If ever anyone is right that there is no future, then that very moment is his last, and what's more is the end of everything. Yet when these philosophers teach that there is no more time to come, they show no trace of terror or despair! When we see them planning and anticipating, we might suspect that they believe in the future as much as anyone else does. (*On the Plurality of Worlds*, p. 207.)

Well of course they do not despair. Even though they deny *that there are future times* they do not deny *that they will continue to exist*, because they do not think that the latter entails the former.[13]

I do not, however, want to say unequivocally that this kind of thing is a mistake. One can have excellent independent reasons to think that a theory's analysis of some notion is just wrong. I, for example, think the branching time theory's analysis of future-indeterminacy is just wrong. And an opponent of the block universe theory might think that its analysis of tense is just wrong (for reasons that parallel common reasons for thinking that David Lewis's analysis of modality is just wrong). If one has those reasons one can use the failure of a theory of time's analysis of tense to argue against its ontology.

---

[12] Tim Maudlin's explanation of his acceptance of the block universe theory's ontology rather than of presentism's echoes this passage ("On the Passing of Time," p. 109).

[13] It is perhaps not crystal clear from the quotation that Lewis is talking about the growing block theory. He asks us to consider philosophers who say that the future is unreal, but if "there are no future times" does not entail "it is false of each thing that it will (continue to) exist" then it will be unclear which someone means to assert when he says that the future is unreal. However, in the quotation Lewis equates "the future is unreal" with "there is no time to come"; this and some things he says after the bit I quoted make it clear what his target is.

6

The Moving Spotlight Theory is Consistent

6.1 The Nuclear Option

We now have several theories of the passage of time on the table. We have the block universe theory with its anemic passage. And we have a couple of versions of the moving spotlight theory with their more robust passage. Which of them, if any, is correct?

Answering a question like this, about opposing theories in some branch of metaphysics, is usually a complicated matter. Each side has arguments that favor its theory. Which way, on balance, do the arguments point? All the arguments have premises. Typically at least some of the arguments on each side are valid arguments with plausible premises. Not all of these premises can be true. So to decide in favor of one of the theories one must say that some appealing claims—the premises of the arguments favoring the other theories—are false. Choosing which to reject is a complicated, messy affair.

Unless there is a nuclear option: an argument against some of the views that is so good, so powerful, that it is decisive all by itself.[1] It renders careful consideration of all the arguments on both sides unnecessary.

Many opponents of objective becoming have thought they had an argument this good. It would have to be an argument with no premises at all, or else an argument with premises so secure that no rational person could reject them. The "gold standard" would be an argument that showed that objective becoming was inconsistent, that there was a contradiction in the very idea that objective becoming occurred. An inconsistent theory cannot be true, so if the moving spotlight theory is inconsistent there is no need to consider how good the arguments in its favor are.

But the moving spotlight theory is not inconsistent. The arguments that have been thought to show that it is do not work.

---

[1] I take the analogy with nuclear warfare from Callender, "The Common Now" (p. 340).

6.2 McTaggart

The most famous argument in the philosophy of time is McTaggart's argument for the unreality of time.[2] Few philosophers think his argument as a whole works. But part of the argument appears to be an argument that the very notion of objective becoming is inconsistent. Philosophical opinion about this part of the argument is polarized. D. H. Mellor wrote that the soundness of this part "seems beyond all reasonable doubt," while Broad called it a "philosophical 'howler'." Dean Zimmerman says that it is McTaggart's "worst argument."[3] I am on the side of those who think that the argument is a complete failure.

To try to say anything about McTaggart's argument is to walk into a dangerous labyrinth. Peter van Inwagen admitted that "There are key sentences in the text in which his argument is presented that, after ten careful readings, are no more than strings of words to me" (*Metaphysics*, p. 81). If van Inwagen cannot understand those sentences then I doubt that I can either. Still, it would be irresponsible for me to ignore the argument completely.

One reason why McTaggart's argument is difficult to understand is that it is difficult to identify just what McTaggart thinks objective becoming would be if it occurred. I am going to bypass this difficulty. Instead of trying to identify the view McTaggart takes himself to be arguing against, I will investigate whether McTaggart's main allegations constitute an argument that either MST-Supertense or MST-Time is inconsistent. A fair bit of what I say in this section is not new, but it is a necessary preamble to the sections that follow.[4]

McTaggart begins by noting, rightly, that

(1) "T is present," "T is past," and "T is future" are incompatible. No time can satisfy more than one of them.[5]

But, McTaggart continues, and here is a central claim of his argument,

(2) If there is objective becoming, then every time is present *and* past *and* future.

---

[2] The argument appears in *The Nature of Existence*, volume 2, sections 329–33. All quotations from McTaggart are from these sections. The argument also appeared in an earlier article called "The Unreality of Time."

[3] See Mellor, *Real Time II* (p. 72); Broad, *Examination of McTaggart's Philosophy*, volume 2, part 1 (p. 316); and Zimmerman, "The A-Theory of Time, The B-Theory of Time, and 'Taking Tense Seriously'" (p. 401).

[4] The secondary literature on McTaggart's argument is enormous. I will not try to provide a comprehensive guide to it. My views about McTaggart's argument are close to, though not exactly the same as, the views Van Cleve expresses in section II of his paper "If Meinong is Wrong, is McTaggart Right?"

[5] McTaggart begins his argument by speaking of events satisfying these predicates, rather than times. But he later speaks of times rather than events and says the same argument applies.

If these claims are true then the idea that there is objective becoming is inconsistent.

So is every time both present and past in MST-Supertense or MST-Time? Of course not. In MST-Supertense exactly one time, call it N, super-is present, full-stop. What is N's relation to pastness? Well in MST-Supertense "T super-is past" is defined to mean "T super-is earlier than the time that super-is present."[6] Since no time super-is earlier than itself, no time, N included, super-is both present and past. Still, some time later than N super-will be present. So N super-will be earlier than the time that super-is present.[7] But this is just to say that N super-will be past. For similar reasons N super-was future.[8]

What about MST-Time? Let T be any time. Then from the perspective of T, T is present. But T is not also past from this perspective. For as in MST-Supertense "T is past" is defined to mean "T is earlier than the time that is present." Of course, from the perspective of a time later than T, T is past. But this does not entail that T is in some unqualified way ("outside of all perspectives") both present and past. It just entails that, from the perspective of T, T is present and will be past.

McTaggart anticipates that a defender of objective becoming will reject (2) and will insist instead that the time that is present will be past. McTaggart's famous reply goes like this:

But what is meant by "has been" and "will be"? . . . When we say that X will be Y, we are asserting X to be Y at a moment of future time. . . . Thus our first statement about M— that it is present, will be past, and has been future—means that M is present at a moment of present time, past at some moment of future time, and future at some moment of past time. But every moment . . . is both past, present, and future. And so a similar difficulty

---

[6] Alternatively, it might be defined to mean "T super-was present." Given the rest of the theory these definitions are equivalent. (McTaggart's argument uses ordinary tense, not supertense. But this distinction is irrelevant here, because in MST-Supertense truth-conditions for tensed sentences about which time is present follow the "change tense to supertense" rule.)

Sometimes it is said that in the moving spotlight theory the properties of presentness, pastness, and futurity are "monadic, intrinsic" properties. This might be taken to preclude either of these definitions of "T super-is past." It might even be taken to entail that "T super-is past" and "T super-is future" are fundamental predicates, lacking definitions in more fundamental terms. If the theory is developed this way then it must still assert that it is a necessary truth that if T super-is past then T super-is earlier than the time that super-is present. But I see no virtue in developing the theory in this way.

[7] There is a potential scope for ambiguity here. I mean "It super-will be the case that (N is earlier than the time that super-is present)."

[8] Savitt, in "A Limited Defense of Passage," points out that there are ways of interpreting "every time is present, and is past, and is future" so that it is true if there is objective becoming. Applied to MST-Supertense his claim is that this sentence is true if we interpret "is" to mean "super-is, super-was, or super-will be." But, Savitt continues, on this interpretation of "is" "T is present" and "T is past" are not incompatible. Having said this, I will stick to the "default" interpretation of "is" as "super-is" in my discussion.

arises. If M is present, there is no moment of past time at which it is past. But the moments of future time, in which it is past, are equally moments of past time, in which it cannot be past . . .

The argument here seems to proceed like this. Suppose that there is objective becoming and let M be the present time. Then M will be past. But "M will be past" just means

(3) There is a moment of time T such that (i) T is future, and (ii) M is past at T.

Furthermore, it is *not* the case that M *was* past. And "it is not the case that M was past" means

(4) There is no time T such that (i) T is past, and (ii) M is past at T.

But, McTaggart continues, (3) and (4) cannot both be true if there is objective becoming, for

(5) If there is objective becoming then every time is present and past and future (so, in particular, every time that is future is also past).

The obvious flaw in this argument is that (5) is just (2) again, the claim that McTaggart's opponent has just rejected! McTaggart is supposed to be giving an argument that the claim that his opponent said was true *in place of* (2) is still inconsistent; it is entirely inappropriate to use (2) itself in the argument.

From the point of view of MST-Supertense there is another flaw in the argument. The argument claims that "M will be past" just means (3). For this argument to even have a chance of being true in MST-Supertense we should read "will be" as "super-will be." Even still the claim is false. "M super-will be past" does not mean "M super-is past at a moment of future time." "M super-is past" is metaphysically complete. It does not need a temporal modifier to make it complete. So if it even makes sense to append "at a moment of future time" to "M super-is past," the resulting sentence has the same truth-value as "M super-is past." Both are false. But "M super-will be past" is true. So it cannot mean the same as (3).

Broad, in his discussion of McTaggart's argument, says that the driving idea behind the argument is reductionism about tense (*Examination of McTaggart's Philosophy*, volume 2, part 1, section 3.11). On Broad's interpretation McTaggart refuses to let "M is present and will be past" stand because it is tensed. McTaggart then produces a tenseless truth-condition for this sentence, and goes on to argue that this truth-condition is inconsistent. (On this interpretation, claim (3) is to be understood as a tenseless truth-condition for "M will be past.") Broad says that

McTaggart's opponent should just reject reductionism about tense, and MST-Supertense embraces this suggestion.[9]

It is perhaps worth noting at this point that a theory is not required to have unreduced supertense to evade McTaggart's argument (as interpreted by Broad). MST-Time makes do with "double" temporal qualifiers rather than supertensed verbs. It gives a truth-condition to "M will be past" that looks a lot like McTaggart's:

"M will be past" is true from the perspective of T if and only if there is a time U later than T such that, from the perspective of U, M is past.

Yet MST-Time is also consistent. So assuming reductionism about tense (narrowly construed) was not McTaggart's only mistake. (I will give a more general characterization of that mistake at the beginning of the next section.)

To conclude my discussion of McTaggart's argument let me say something about D. H. Mellor's reconstruction of it.[10] His version of the argument is not as blatantly question-begging as mine. Nor does Mellor seem to just assume that reductionism about tense is true. But his reconstruction of the argument still fails. Put in dialogue form Mellor's version, addressed to MST-Supertense, looks like this:

M: The idea that there really is such a thing as objective becoming is inconsistent.

O: It seems perfectly consistent to me. What do you think is wrong with it?

M: Well if there is objective becoming then each time super-is past, present, and future. But nothing can be all three.

O: No, that's not true at all. If there is objective becoming then one time super-is present, super-was future, and super-will be past; other times super-are past, super-were present, and super-were future. And so on.

M: So the thing you say is true is not a sentence containing the "simple" predicates "T super-is past," "T super-is future," and so on, but a sentence containing more complicated predicates like "T super-will be past" and "T

---

[9] In his earlier paper on the unreality of time McTaggart wrote, "Our ground for rejecting time, it may be said, is that time cannot be explained without assuming time. But may this not prove—not that time is invalid, but rather that time is ultimate? It is impossible to explain, for example, goodness or truth unless by bringing in the term to be explained. . . . But we do not therefore reject the notion as erroneous, but accept it as something ultimate, which, while it does not admit of explanation, does not require it" (p. 470). This is the believer in MST-Supertense's response exactly. McTaggart goes on to explain why he thinks it is wrong, but I do not understand what he says.

[10] It is in chapter 7 of *Real Time II*. A similar reconstruction is in Dummett, "A Defense of McTaggart's Proof of the Unreality of Time."

super-was future." Using these predicates may appear to save you but really they land you back in the hot water you are trying to escape. For if there is objective becoming then each time *super-was* past and *super-will be* future. But nothing can be both.

O: Wait, the existence of objective becoming does not require that! Let T be a time that super-was past. (Any time sufficiently far in the past is a time that super-was past.) It's just false that T super-will be future. What's true is that *it super-was the case that* T super-will be future.

M: But there is still inconsistency when we use yet more complex predicates like "it super-was the case that (T super-will be future)."[11] If there is objective becoming then each time T is such that (i) it super-was the case that (T super-was past) and (ii) it super-will be the case that (T super-will be future). But (once again) nothing can be both.

O: . . .

This dialogue should not convince anyone that the claim that there is objective becoming is inconsistent. At each stage M accuses objective becoming of being inconsistent and O shows that allegation to be false. And even an infinite sequence of false allegations does not add up to a good argument.

This complaint about McTaggart's argument goes back at least to Prior, whom I cannot resist quoting:

We are presented, to begin with . . . , with a statement which is plainly wrong. . . . This is corrected to something which is plainly right. . . . This is then expanded . . . to something which, in the meaning intended, is wrong. It is then corrected to something a little more complicated which is right. . . . we shall have to go on *ad infinitum*. Even if we are some-how compelled to move forward in this way, we only get contradictions half the time, and it is not obvious why we should regard these rather than their running mates as the correct stopping-points. (*Past, Present and Future*, pp. 5–6.)

Prior is too kind in this last sentence; it is not just that it is "not obvious" that M's allegations are correct, but (as he says earlier) that they are "clearly wrong."[12] But let's get back to Mellor. Mellor says, about the potentially infinite sequence of exchanges between M and O, "we have an endless regress, a regress that is vicious because at no stage in it can all the A-facts it entails be consistently stated" (*Real Time II*, p. 74). (By "A-facts" we can take him to mean supertensed

---

[11] In his presentation Mellor focuses on tenseless predicates like "((T is future) at a future time) at a past time." But he also says it makes no difference whether we focus on these or on predicates containing iterated tense operators.

[12] I also like Van Cleve's way of putting the point: "it is not the defender of time who continually repeats a hopeless maneuver but rather Mctaggart who continually repeats a baseless charge" ("If Meinong is Wrong, is McTaggart Right?," p. 237).

sentences about which instants are past, present, and future.) But O never says anything inconsistent. Say that sentences of the form "T super-is present," "T super-is past," and "T super-is future" are "simple sentences." At stage n of the dialogue O makes statements in which a simple sentence is prefixed by no more than n supertense operators. Then at stage $n + 1$ M says that further statements with n prefixed supertense operators are true if there is objective becoming; but that those statements are inconsistent. O denies that those statements are true if there is objective becoming and says that instead what are true are certain statements in which a simple sentence is prefixed by no more than $n + 1$ supertense operators. The truth behind Mellor's claim that "at no stage can all the A-facts be consistently stated" is this: if O is limited to making statements in which a simple sentence is prefixed by no more than some fixed number n of supertense operators, then O cannot state all the truths in his theory. Some of those truths have more than n supertense operators. But this is not a flaw in the theory.

6.3 McTaggart Redux: Smith

Some philosophers think that McTaggart's argument, despite its obscurity, is basically right. They set out to eliminate the obscurity. I will discuss three such arguments and show that they do not succeed in showing either MST-Supertense or MST-Time to be inconsistent. Seeing why they fail provides insight, I think, into how these theories work.

   These arguments may avoid some of the problems McTaggart's faces, but in the end they make the same basic mistake McTaggart made. Broad said that one of McTaggart's mistakes was to assume reductionism about tense, and thereby beg the question against MST-Supertense. But this is an instance of a more general mistake. In the moving spotlight theory there is a sense in which time is two-dimensional. Time is literally two-dimensional in MST-Supertime. In MST-Supertense and MST-Time it is not, but there are still two dimensions of "temporal talk." In MST-Supertense supertense is primitive, so what is left of the two-dimensionality of time is the fact that we can talk both about what super-is the case at later times and also about what super-will be the case at the time that super-is present. In MST-Time supertense is not primitive; what is left of the two-dimensionality of time is the distinction between how things vary with time *from a single perspective* and how the way things are *at a single time* varies *across* perspectives. From a single perspective I am in different places at different times; while from one perspective but not from others a given time is present. By assuming reductionism about tense McTaggart in effect tries to force the two dimensions of temporal talk in the moving spotlight theory into a single dimension. No wonder he thought the theory was inconsistent. The arguments I will discuss now fail for similar reasons.

This mistake is most obvious in Nicholas J. J. Smith's attempt to rehabilitate McTaggart's argument.[13] The theory he directs his argument against resembles MST-Time closely enough that I will speak as if this theory is his target. Smith announces that he will defend McTaggart's allegation that in the theory each time is past, present, and future. Smith begins by saying, correctly, that MST-Time associates different pictures of reality with different times—as in Figure 6.1.

Then Smith states his key premise:

if these different diagrams are meant to represent the situation at different *normal* times then they *must* be able to be combined into one four-dimensional spacetime diagram. The A-theorist [or defender of MST-Time] cannot draw one diagram showing the situation as of 1800, and another one showing the situation as of 1900—where 1800 and 1900 are two normal times (as opposed to hypertimes)—and then maintain that the two diagrams cannot be combined into one: that would just be to misunderstand spacetime diagrams. . . . It is simply a confusion to think *both* that "nothing that happens happens outside these four walls of spacetime—there is no other time dimension" *and* that one cannot draw a single four-dimensional spacetime diagram representing the entirety of what has happened, is happening and will happen. But then of course—and this is my point—when we do combine the pictures into a single spacetime diagram, we instantly see the contradiction in the A-theory: every time has to be shown as having all three of the incompatible A-properties *past, present,* and *future.* (P. 241.)

The pictures in Figure 6.1 show how time is from the perspectives of two different times. Smith's premise is that there must be a single picture that shows how time is full-stop, or how time is from a perspective outside of time. And in MST-Time there is not. There is no such thing as how reality is "full-stop" in MST-Time, and there is no perspective outside of time. Force such a picture onto the theory and it does become inconsistent.

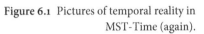
Figure 6.1 Pictures of temporal reality in MST-Time (again).

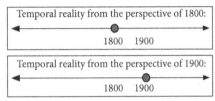

---

[13] It appears in his paper "Inconsistency in the A-theory." Smith's is the most recent of the three arguments I will discuss. I discuss it first because it is more similar to McTaggart's than the others.

So Smith only gets a contradiction from MST-Time with the help of his premise.[14] Thus he only shows the theory to be inconsistent if his premise is a logical truth. And while I agree that his premise is true I do not think it is a logical truth. Smith himself says that rejecting his premise betrays a misunderstanding of spacetime diagrams, but I do not see why. (The diagrams in Figure 6.1 are time diagrams, not spacetime diagrams, but the difference is not important.)

Smith's premise is of course true in the block universe. In that theory, the various ways that reality "looks" from the perspectives of various times can be combined into a single way that reality "looks" from some atemporal perspective. But, again, I do not think that a theory needs to be this way to be consistent.

By insisting that there be one single time diagram that shows how temporal reality is Smith tries to force the two dimensions of temporal talk in the moving spotlight theory into a single dimension. But a theory with two dimensions of temporal talk needs more than a single diagram with (normal, one-dimensional) time in it to completely represent temporal reality.

6.4 McTaggart Redux: Mellor

After Mellor presents his reconstruction of McTaggart's argument he advances another argument that, he claims, arrives at "McTaggart's contradiction, expressed in a way that allows no regress and hence no riposte" (p. 79).[15] Suppose that in 2003 some very confused soul said "Noon, January 1st, 2004 is past." And suppose that in 2005 someone else, less confused, said these same words. Then we have two tokens of the same sentence-type K, uttered at different times. A picture of what is going on if the block universe theory is true is in Figure 6.2.

The basic idea behind Mellor's argument seems to be that if there is objective becoming then sentence tokens change their truth-value as time passes, which is impossible. McTaggart thought he had found a contradiction in the idea that times change from being future to present to past; Mellor thinks he has found a contradiction in the idea that sentence tokens change from being true to false (or vice versa).

---

[14] Other philosophers have identified similar claims as premises of McTaggartian arguments. Horwich, for example, puts it as the assumption that there is a "time-neutral body of absolute facts" (*Asymmetries in Time*, p. 27).

[15] I focus on the argument in chapter 7 of *Real Time II*. My presentation makes some inessential modifications to Mellor's argument. Mellor presented an earlier version in *Real Time* that generated a lot of discussion; two useful references on it are Priest, "Tense and Truth-Conditions," and Percival, "A Presentist's Refutation of Mellor's McTaggart."

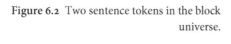

Figure 6.2 Two sentence tokens in the block universe.

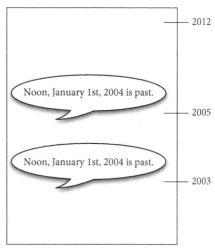

In my reconstruction Mellor's argument uses these four claims:

(1) The 2003 sentence token has the same truth-value at every time; similarly for the 2005 sentence token.
(2) For any time T, all tokens of sentence-type K have the same truth-value at T.
(3) The 2003 sentence token is false at (any time in) 2003.
(4) The 2005 sentence token is true at (any time in) 2005.

So, the argument is that if there is objective becoming then (1) through (4) are all true. But in fact they are inconsistent. For by (3) the 2003 sentence token is false in 2003. Then by (1) that sentence token is also false in 2005. But then by (2) the 2005 sentence token is false in 2005. But by (4) this is not so.

There is something strange going on in Mellor's argument. We are used to the idea that a sentence type may have a truth-value only relative to a time. But (1) through (4) speak of sentence tokens having truth-values at times. At least some believers in the block universe will find it strange to talk about sentence tokens having truth-values relative to times at all. They will say, for example, that the 2005 sentence token is just plain true.[16] But this thought is not much help to

---

[16] This can all be made more intelligible if we indulge briefly in some proposition-talk. If there are such things as propositions then they are the primary bearers of truth-values. Each sentence token expresses a proposition, and the truth-value of the token is determined by the truth-value of the associated proposition. Now some believers in the block universe theory think that truth for propositions is not time-relative. This goes naturally with the idea that truth for sentence tokens is not time-relative either. Other believers in the block universe theory favor the view that propositional truth is time-relative. They will have no problem with talk of sentence tokens being true or false at times.

a defender of objective becoming. For MST-Time, at least, does say that sentence tokens have truth-values relative to times.

So is (1) true if there is objective becoming? And what about (2)? Mellor thinks that (1) is obvious, and that everyone, whether they believe in objective becoming or not, accepts it: "No one thinks, for example, that my death will posthumously verify every premature announcement of it!" (pp. 78–9). His rationale for (2)'s being true if there is objective becoming is a little murkier. Here is the argument he gives on p. 78, adapted to my example:

if the 2003 and 2005 sentence tokens are both made true by the A-fact that noon, January 1st, 2004 is past, they must both be true when this *is* a fact and false when it is not. So in 2003, when noon, January 1st, 2004 is still future, the two sentence tokens must both be false; and in 2005, when it is past, they must both be true.

What does this mean? Presumably an A-fact is a fact statable by a true sentence containing tense or A-vocabulary. And presumably a sentence token is "made true" by a fact if some true sentence expressing that fact is the truth-condition for that sentence. With these definitions does Mellor's defense of (2) work?

Before answering it is worth backing up and working through how Mellor's argument looks with some version of the moving spotlight theory explicitly in mind.[17] Doing this for MST-Supertense is going to be tricky, because that theory denies that even the sentence type "T is past" has its truth-value relative to times. So let's start with MST-Time.

If we are to read Mellor's argument as directed against MST-Time then we should understand talk of a sentence token's being true at a time T as talk about a token's being true from the perspective of T. If we make this explicit the argument looks like this:

(1) The 2003 sentence token has the same truth-value from the perspective of every time; similarly for the 2005 sentence token.
(2) For any time T, all tokens of sentence-type K have the same truth-value from the perspective of T.
(3) The 2003 sentence token is false from the perspective of any time in 2003.
(4) The 2005 sentence token is true from the perspective of any time in 2005.

We also need a picture of the scenario under discussion that assumes MST-Time rather than the block universe theory. Figure 6.3 is what we need. It shows

---

[17] Percival, in "A Presentist's Refutation of Mellor's McTaggart," argues that Mellor's argument does not work against presentism. But it would be bad if presentism were the only alternative to the block universe of which this were true, since there is no robust passage in presentism.

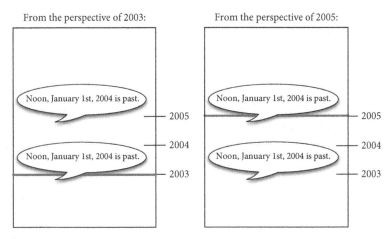

Figure 6.3 Two sentence tokens in MST-Time.

how that scenario is from the perspective of 2003 and also how it is from the perspective of 2005.

Figure 6.3 shows clearly where Mellor's argument goes wrong. Both sentence tokens are true from the perspective of 2005 and both false from the perspective of 2003. So (2) is true. Furthermore, (3) and (4) are also true. For the 2003 token is false in the left panel of Figure 6.3 and the 2005 token is true in the right one. But (1) is false. The 2003 token is false from the perspective of 2003, true from the perspective of 2005.

(As an aside, Mellor's murky rationale for (2) now seems less murky. From the perspective of 2005 "it is a fact that" 2004 is past. But "when this is a fact"— from the perspective of a time from which this is true—both sentence tokens must be true. So both sentence tokens must be true from the perspective of 2005. Similarly, since from the perspective of 2003 it is not a fact that 2004 is past, both sentence tokens are false from this perspective.)

So what of Mellor's argument that (1) is obvious? "No one thinks that my death will posthumously verify every premature announcement of it!" Or, to adapt his complaint to our example, suppose I prematurely announce "3000 is present." Mellor might complain: "No one thinks that presentness's arrival at 3000 (so to speak) will *ex post facto* verify that announcement!" Is that right? Maybe people who already accept the block universe theory will agree. But I see no independent reason to agree with Mellor's complaint. Still, there is a related complaint one might make:

Someone would be *making a mistake* if he announced (now) that 3000 is present. But if that announcement is true from *some* perspective then where is the mistake?

But I do not think this complaint goes anywhere. He would be making a mistake because his announcement would be false from the perspective of the time he made it.

So much for Mellor's argument and MST-Time. How does it fare against MST-Supertense? In a way the argument fails to engage with that theory. As I have said before, MST-Supertense says that "M is past" is metaphysically complete. Adding "at T" is idle. "M is past at T" has the same truth-value no matter what time is put in for "T." But Mellor's argument is full of these idle temporal qualifiers. Here are the four claims again in their original form:

(1) The 2003 sentence token has the same truth-value at every time; similarly for the 2005 sentence token.
(2) For any time T, all tokens of sentence-type K have the same truth-value at T.
(3) The 2003 sentence token is false at (any time in) 2003.
(4) The 2005 sentence token is true at (any time in) 2005.

It seems that insofar as (1) makes sense in MST-Supertense it is true. Claim (2) is also true, since every token of "Noon, January 1st, 2004 is past" is just plain true. But then (3) is obviously false, so the argument fails.

But maybe this response is not very fair to the argument. Let's try to tailor the argument so that it applies more directly to MST-Supertense. A natural way to revise the premises is to eliminate the talk of time-relative truth values and instead talk of supertensed truth-values:

(1-2) Whatever truth-value the 2003 sentence token super-has, it super-will always have and super-has always had that truth-value. Similarly for the 2005 sentence token.
(2-2) It super-is always the case that (all tokens of sentence-type K super-have the same truth-value).
(3-2) The 2003 sentence token super-was false (nine "superyears" ago).
(4-2) The 2005 sentence token super-was true (seven "superyears" ago).

Figure 6.4 is a picture of the scenario these claims are about that is adapted to MST-Supertense.

This version of Mellor's argument also fails. Claim (1-2) is false, for reasons similar to the reasons why (1) in the argument against MST-Time is false. It is part of the theory that tokens of "T is past" do not super-always have the same truth-value. While the 2005 sentence token super-was false nine "superyears" ago, it super-is true now.

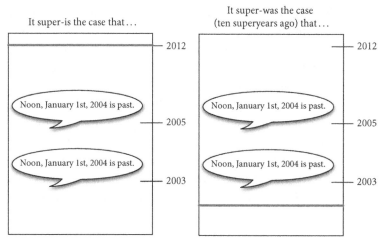

Figure 6.4 Two sentence tokens in MST-Supertense.

6.5 McTaggart Redux: Van Cleve

James Van Cleve, in "If Meinong is Wrong, is McTaggart Right?", presents an argument that, he says, is "very much like McTaggart's in substance and strategy" (pp. 239–40). The argument begins with some premises that seem so innocent that no one would reject them. Then at the end we arrive at the claim that the moving spotlight theory is inconsistent given those premises.

Let's start along Van Cleve's garden path and see where it leads. His first premise is this:[18]

(1) For all times T, if X precedes Y at T then X exists at T and Y exists at T.

The variables X and Y in (1) range over events as well as instants of time. Now I do not think that, fundamentally, there are such things as events. So I think that any truth that can be expressed using event-talk can also be expressed without it. (It may take a much more complicated sentence in a much more powerful language.) But I am not going to try to translate Van Cleve's argument into a language that does without event-talk. It is not necessary to do that to see its flaws. Still, we should approach the event-talk in his argument with suspicion, and make sure we understand exactly what assumptions about events he is making.

So let me say a few things about events (assuming that they exist, if not in a fundamental sense then in some less fundamental sense). If George climbs a tree

---

[18] In what follows I have simplified Van Cleve's argument in a few ways. I have left out some inessential steps and have made minor modifications to a couple of premises.

then, we usually say, there is an event, George's climb, that occurs just as long as George is climbing and occurs in just the places George is while he climbs. But that is an easy case. What about a room that is empty from noon to 1pm? Are there any events going on in that room? Van Cleve's argument is easiest to discuss if we assume a liberal view about event existence. Let us assume that even in a region of space that is completely empty during some stretch of time there is an event that goes on just in that region just during that stretch of time.

Now that we have some clarity about events I want to say something about the word "precedes" that appears in (1). When applied to instants of time "X precedes Y" just means "X is earlier than Y." When applied to events it means something similar. "E precedes F" means "the time at which E occurs is earlier than the time at which F occurs."[19]

Another thing to notice about (1) is that again we have temporal modifiers ("at some time," "at every time") that a proponent of MST-Supertense, and of the block universe theory, will find odd. "Noon (on a particular day) precedes 1 pm full-stop, not 'at some time' or 'at every time'" they may want to complain. But there is nothing odd about these temporal modifiers in MST-Time. On that theory the only perspectives on reality are perspectives from particular times, so no sentence is metaphysically complete without a temporal modifier in it. In light of this the most natural thing to do when discussing Van Cleve's argument is to take it to be aimed at MST-Time. (As with McTaggart and Mellor it is not clear whether Van Cleve aims his argument at MST-Time, at MST-Supertense, at some other theory of objective becoming, or collectively at all of them.)

After all this hand-wringing does (1) look plausible? Or should we already be suspicious? Here it is again:

(1) For all times T, if X precedes Y at T then X exists at T and Y exists at T.

It does still look plausible, and in fact I think that believers in MST-Time should think that (1) is true.

What's next? At this point Van Cleve makes the following assumption:

If X (an event or a time) is present then there is a corresponding event. Let us call this event "X's possession of presentness," or "N[X]" for short. The same goes if X is past (P[X]) and if X is future (F[X]).[20]

---

[19] If E and F are events that last for more than one instant then it is trickier to say what it takes for E to precede F. Maybe it is enough for the time at which E starts to be earlier than the time at which F starts. But we will not have to worry about this case.

[20] An event is present if and only if one of the times at which it occurs is present.

This assumption seems safe given the liberalism about event existence we are assuming.[21] Here is Van Cleve's next premise:

(2) If there is objective becoming, then if X precedes Y, N[X] precedes N[Y].

That looks right. Presentness "moves along the series of times" from earlier times to later times. So its arrival at an earlier time occurs earlier than its arrival at a later time.

He says something similar is true for pastness:

(3) If there is objective becoming, then if X precedes Y, P[X] precedes P[Y].

Now Van Cleve believes he has set his trap, and he invites believers in objective becoming to walk in. Believers in the moving spotlight theory will certainly agree that

(4) If there is objective becoming, then there are times R and S such that R precedes S.

Aha! (4) combines with (2) and (3) to give

(5) If there is objective becoming then N[R] precedes N[S] and P[R] precedes P[S].

But now by (1) we can conclude

(6) If there is objective becoming then N[R] and P[R] exist at each time.

Van Cleve's final premise is

(7) If N[X] exists at T then X is present at T, and if P[X] exists at T then X is past at T.

Surely if the event of X's having presentness exists at T then X is present at T! And surely if the event of X's having pastness exists at T then X is past at T! But (6) said that for a particular time R, N[R] and P[R] exist *at every time*. So (7) and (6) entail that objective becoming is inconsistent:

(8) If there is objective becoming then R is both present and past at every time.

---

[21] There may be grounds for doubting this assumption, even if we assume liberalism. Where and when does this event N[X] occur? Even if this is a problem I want to focus on deeper and more interesting flaws in the argument.

There are several things wrong with this argument. The more straightforward problem (it will lead to the others) is that it is not formally valid. Conclusion (6) does not follow from (5) and (1). Look at them again:

(1) For all times T, if X precedes Y at T then X exists at T and Y exists at T.
(5) If there is objective becoming then N[R] precedes N[S] and P[R] precedes [PS].
(6) If there is objective becoming then N[R] and P[R] exist at each time.

There is no "at some time" or "at T" in (5). Once we notice this we see that there are not any in (2), (3), or (4) either. They need to be there for the argument as a whole to be valid.

Let's have another look at the argument as a whole, this time with all the "at T"s put in:

(1) For all X and Y and for all times T, if X precedes Y at T then X exists at T and Y exists at T.
(2) If there is objective becoming, then if X precedes Y *at some time T*, N[X] precedes N[Y] *at T*.
(3) If there is objective becoming, then if X precedes Y *at some time T*, P[X] precedes P[Y] *at T*.
(4) If there is objective becoming, then there are times R and S such that R precedes S *at each time.*
(5) If there is objective becoming then N[R] precedes N[S] *at each time* and P[R] precedes P[S] *at each time* (from (2)–(4)).
(6) If there is objective becoming then N[R] and P[R] exist at each time (from (5) and (1)).
(7) If N[X] exists at T then X is present at T, and if P[X] exists at T then X is past at T.
(8) If there is objective becoming then R is both present and past at each time (from (6) and (7)).

Now we can see the main flaw in this argument. It confuses two meanings that "event E exists at T" has in MST-Time. One meaning—the more ordinary meaning, the only meaning it has in the block universe theory—is "event E is temporally located at T," or "T is among the times at which E occurs." The other meaning, the one unique to MST-Time, is "from the perspective of T, E exists." And more generally sometimes "S at T" means, in MST-Time, "from the perspective of T, S." So which meaning does "at T" have in each of Van Cleve's premises?

There is no way to interpret the "at T"s in Van Cleve's argument so that it is valid and has true premises. The easiest way to see this is to focus on (2)—one of the premises from which Van Cleve omitted the temporal qualifiers in his statement of the argument:

(2) If there is objective becoming, then if X precedes Y at some time T, N[X] precedes N[Y] at T.

Statement (2) is false if "at T" means "from the perspective of time T." Let U be some time and V a later time. So U precedes V—from the perspective of every time (facts about which times are earlier than which do not vary from one perspective to another). But there is no perspective from which N[U] precedes N[V]. That is because there is no perspective from which both these events exist, and, given (1), both have to exist in order for one to precede the other.

Why is there no perspective from which both events exist? The event N[X] exists from some perspective if and only if X is present from that perspective.[22] Thus from the perspective of U, N[U] exists and N[V] does not.

Van Cleve's paper is titled "If Meinong is Wrong, is McTaggart Right?" Premise (1) is his "anti-Meinongian" premise, and he suggests that moving spotlight theorists who want to save their theory from inconsistency will have to "go Meinongian" and reject it. But as I see it a defender of MST-Time should accept (1) and use it as part of his grounds for rejecting (2).[23]

I have identified (2) as one of the bad premises. I did say something in its favor when I presented it:

(2) looks right: presentness "moves along the series of times" from earlier times to later times. So its arrival at an earlier time occurs earlier than its arrival at a later time.

This is sleight-of-hand. The sentence following "So," obvious-sounding as it is, is not correct. N[U] does not precede N[V]. What is true is this: *the perspective from which N[U] exists* precedes the perspective from which N[V] exists.

I have said that I prefer to avoid event-talk. Van Cleve's argument illustrates one of the dangers of it. I said earlier that in MST-Time the only thing that changes from one perspective to another is which time is present. But if there

---

[22] At least, this seems like the most plausible thing to say in MST-Time about what it takes for N[X] to exist. Alternative views are possible. But they are no help to Van Cleve. For if one says that N[X] can exist from some perspective even though X is not present from that perspective then premise (7) is false.

[23] I have been interpreting "at T" as it occurs in (1) to mean "from the perspective of time T." That is the only interpretation that makes (1) true. On the other interpretation (1) says that if event E precedes event F from the perspective of a time T then E *occurs at* T and F *occurs at* T. That is obviously false. An event can exist from the perspective of some time without occurring at that time (relative to that perspective). The big bang exists from the perspective of 2012.

are strange events like N[T] then this is not true. If there are such events then there are other differences between perspectives. There are differences in what exists: the event N[T] exists only from the perspective of time T.

I have one more comment about event-talk and the passage of time. Smart spends a great deal of "The River of Time" drawing attention to (what he takes to be) confusions surrounding event-talk. This is partly because early discussions of the moving spotlight theory made essential use of event-talk. McTaggart started it all by characterizing the passage of time as the process whereby events that start out future become first present and then past. But attending carefully to how event-talk works cannot show the moving spotlight theory to be inconsistent, because my presentations of the various versions of the theory do not use event-talk at all.

7

How Fast Does Time Pass?

7.1 An Unanswerable Question?

The nuclear strategy is a failure. McTaggart's argument does not work, and, what is worse, is confusing and hard to follow. Even if it were sound it would be reasonable to doubt that it succeeded.

Some opponents of objective becoming think they have another, better, weapon in their arsenal. Instead of advancing an argument filled with technical philosophical notions they ask a simple question:

How fast does time pass?

The believer in objective becoming, they argue, cannot give an adequate answer to this question.

As with so many of the great ideas in the philosophy of time, it was, to my knowledge, C. D. Broad who first pursued this one. It is the first of several "fatal objections" he raises in his 1938 discussion of the moving spotlight theory:

If anything moves, it must move with some determinate velocity. It will always be sensible to ask "How fast does it move?" even if we have no means of answering the question. Now this is equivalent to asking "How great a distance will it have traversed in unit time-lapse?" But here the series along which presentness is supposed to move is temporal and not spatial. In it "distance" *is* time-lapse. So the question becomes "How great a time-lapse will presentness have traversed in unit time-lapse?" And this question seems to be meaningless. (*Examination of McTaggart's Philosophy*, volume 2, part 1, p. 277.)

Broad's argument here seems to be this:

(1) If something moves, it moves with some determinate velocity. The question "How fast does it move?" is a meaningful question.

(2) If the moving spotlight theory is true then presentness moves (along the series of times).

(3) But "How fast does presentness move along the series of times?," which just means "How great a time-lapse will presentness have traversed in unit time-lapse?," is a meaningless question.

(C) Therefore, the moving spotlight theory is false.

Eleven years later, in his 1949 paper "The River of Time," Smart also argued that considering the question "How fast does time pass?" shows the moving spotlight theory to be false. While he started in the same place Broad did, he ended up somewhere different:

with respect to motion in space it is always possible to ask "how fast is it?". An express train, for example, may be moving at 88 feet per second. The question "How fast is it moving?" is a sensible question with a definite answer: "88 feet per second". We may not in fact know the answer, but we do at any rate know what sort of answer is required. Contrast the pseudo-question "how fast am I advancing through time?" or "How fast did time flow yesterday?". We do not know how we ought to set about answering it. What sort of measurements ought we to make? We do not even know the sort of units in which our answer should be expressed. "I am advancing through time at how many seconds per—?" we might begin, and then we should have to stop. What could possibly fill the blank? Not "seconds" surely. In that case the most we could hope for would be the not very illuminating remark that there is just one second in every second.

It is clear, then, that we cannot talk about time as a river, about the flow of time, . . . without being in great danger of falling into absurdity. (p. 485)

When Smart calls "How fast did time flow yesterday?" a "pseudo-question" and asks "What sort of measurements ought we to make?" he invokes the specter of logical positivism. The positivists' infamous verifiability criterion of mean-ingfulness said, more or less, that a statement is meaningless unless its truth or falsity can be verified by some experiment with observationally distinguish-able outcomes. Smart does not answer his question "What sort of measurements ought we to make?" but the implied answer is that there are no measurements one could make to discover how fast time passes. So without explicitly endorsing verificationism Smart seems here to be defending Broad's charge (3) that "How fast does time pass?" is meaningless.[1] (Broad himself gave no defense.)

But then Smart adds something else. He entertains the possibility that the mov-ing spotlight theorist will say that time passes at one second per second. What is wrong with that answer? His response seems to be

---

[1] It is not a great idea to appeal to verificationism in an argument for the block universe theory, as the positivists thought their doctrine showed all metaphysical theses, including the block universe theory, to be meaningless.

(4) "Time passes at one second per second" just means "there is just one second in every second."[2]

But, Smart says, "there is just one second in every second" is "not very illuminating."

Just what argument against the moving spotlight theory could come from (4)? Is it bad if the moving spotlight theory's answer to "How fast does time pass?" is unilluminating? What is the connection between how illuminating the answer is, and whether the answer is right?

Admittedly, Smart's goal is to show that talk of time flowing like a river can lead to philosophical confusion and even "absurdity." That goal is more modest than the goal of showing the moving spotlight theory to be false. Other philosophers who aim at this second goal have sharpened (4) into a more useful weapon. Schlesinger, who defends the moving spotlight theory, points the way to an argument when he writes, "to say that time flows at the rate of one minute per minute does not seem to convey any information about the speed of time" ("How Time Flies," p. 509). This remark suggests building the following argument on (4):

(4) "Time passes at one second per second" just means "there is just one second in every second."

(5) If the moving spotlight theory is true then there must be something informative to say about the rate of time's (robust) passage. A claim about the rate of time's passage that is true by definition cannot be the correct answer to this question.

(6) But "There is just one second in every second" is true by definition.

(7) And if the moving spotlight theory is true then "one second per second" is the only potential answer to "How fast does time pass?" If the theory cannot give that answer it can give no answer.

(C) Therefore, the moving spotlight theory is false.

So far meditating on the question "How fast does time pass?" has produced two arguments against the moving spotlight theory. Further meditation will produce more. But I want to pause now and explain what is wrong with these two.

7.2 Changes without Rates of Change

Broad goes wrong right from the beginning. Look again at his first two premises:

[2] In the quoted passage it is "I am advancing through time at one second per second" that Smart equates with "there is just one second in every second." But context strongly suggests he would say the same about "Time passes at one second per second."

(1) If something moves, it moves with some determinate velocity. The question "How fast does it move?" is a meaningful question.

(2) If the moving spotlight theory is true then presentness moves (along the series of times).

First a quibble. This argument presupposes that in the moving spotlight theory, something—presentness—moves. Earlier I presented a nominalistic version of the theory in which there is no such thing as presentness. In the theory time changes, but it does not change by moving or flowing.

This is, again, a quibble. Some defenders of the moving spotlight might be happy saying that in their theory there really is such a thing as presentness that moves along the series of times. But this quibble leads somewhere important. Is the analogue of (1) true about all change? Is it true that if something changes, it must change at some definite rate? Certainly not. Consider George. During his adventures in the city George got hungrier and hungrier. Must he have been getting hungrier at some determinate rate? Is it sensible to ask, precisely how fast was his hunger increasing? No.

There is no precise rate at which hunger increases because hunger is not a measurable quantity, and it is only sensible to ask about the precise rate at which some feature is changing if that feature is a measurable quantity. I will say more about what measurable quantities are in Section 7.4. For now this observation is enough. While it may make sense to say that Joe is hungrier than Moe, it makes no sense to say that Joe is twice as hungry as Moe, or any particular number of times as hungry as Moe. But that sort of thing needs to make sense for a feature to be a measurable quantity.

Why is being a measurable quantity necessary for having a precise rate of change? If it makes no sense to ask how much hungrier Joe is than Moe, then it also makes no sense to ask how much hungrier George is now than he was an hour ago. But he is getting hungrier at a definite rate only if this question makes sense.[3]

If you think that it makes sense to say that Joe is twice as hungry as Moe then just change the example. Take hardness. The most common scale for measuring hardness, the Mohs scale, assigns substance S a higher number than substance R

---

[3] Even if it makes no sense to say that George is twice as hungry as he was an hour ago, it might still make sense to say that he is much hungrier than he was an hour ago. Then perhaps it makes sense to say that he got hungrier very fast, even if there is no way to assign a number to the rate at which he got hungrier. Similarly, if over the last hour both George and Fred got extremely hungry— as hungry as they could be—it might make sense to say that George got hungrier faster than Fred did. (Suppose George reached his state of maximum hunger earlier than Fred did.) That is still not enough for there to be a way to assign a number to the rate at which George or Fred got hungrier.

when S can scratch R but R cannot scratch S. If this scale reveals all there is to hardness then there is no sense to the question whether S is twice as hard as R.[4]

The moving spotlight theorist can use these ideas to reject Broad's premise (1). She can say that Smart was putting his thumb on the scale when he compared the rate of time's passage to the rate of a train's motion. It should be compared instead to the rate at which George is getting hungrier.[5] But the analogy is not quite perfect, so more needs to be said. Speaking quite generally, a rate of change is got by dividing the amount of the relevant change by the amount of time it took for the change to occur. The reason that it makes no sense to ask how fast George is getting hungrier is that there is no precise amount of the relevant change that occurred, no precise amount by which George's hunger level has changed, during the last minute. But when it comes to the motion of presentness there is a precise amount of the relevant change. The relevant change is how far presentness has moved. It has moved some number of seconds. What the moving spotlight theorist can deny is that there is some definite "amount of time" it took for the change to occur. I put "amount of time" in quotes because, in MST-Supertense at least, the rate of time's passage is not a matter of how much time it took presentness to travel a certain distance in time, but a matter of how much supertime it took. Or, since the theory substitutes tense operators for talk of supertime, it is a matter of how long it super-took for presentness to travel that distance. This is, in fact, exactly what MST-Supertense, as so far formulated, already says. The theory says that one particular time, N, super-is present. And the time that is one second after N super-will be present. But the theory so far says nothing about how long it super-will be until that happens.

I have so far not said just what statements in MST-Supertense collectively characterize the robust passage of time. All I have said is that each time later than the

---

[4] The Mohs scale is not the only scale for measuring hardness. The Vickers scale appears to make hardness a measurable quantity. But it is not clear to me whether the Vickers scale and the Mohs scale are two scales for measuring the very same quantity, or instead whether they are scales for measuring distinct but similar quantities, quantities which are equally good precise candidates for what we mean by the vague word "hardness." We can sidestep these issues by just stipulating that "X is harder than Y" means, here anyway, "X can scratch Y but Y cannot scratch Y," and that this is the only notion of hardness we are working with. Given this stipulation there is no sense to be made of the question whether one thing is twice as hard as another.

[5] In his well-known paper "How Fast Does Time Pass?" Ned Markosian also suggests that a believer in objective becoming may say that it makes no sense to ask how fast time passes. But the grounds he offers are different. He suggests that to ask this is to make a category mistake. He draws an analogy with Wittgenstein's claim that it makes no sense to ask how long the standard meter is.

I do not find this suggestion helpful because Wittgenstein is wrong. It makes perfect sense to ask how long the standard meter is. (Markosian himself does not endorse this response to "How fast does time pass?"; he is agnostic between it and two other responses, one of which I will discuss in Section 7.5.)

present time super-will be present, and each time earlier than the present time super-was present. But these statements alone are inadequate.

One way to see that they are inadequate is to look at what these state-ments do and do not rule out when they are translated into the language of MST-Supertime. To translate we need a translation manual, which I will now provide. (The translation rules are closely connected to the truth-conditions for supertensed talk in MST-Supertime.) For any supertensed sentence P and term X denoting a point in supertime the language of MST-Supertime is to be definitionally extended to include the sentence $\ulcorner(P)^X\urcorner$. The idea is that $\ulcorner(P)^X\urcorner$ is true if and only if P is true when its supertense operators are interpreted un-der the assumption that the supertime denoted by X is "superpresent." Let S be a name of some one particular supertime; then the translation of any supertensed sentence P is $\ulcorner(P)^S\urcorner$. The definition of $\ulcorner(P)^X\urcorner$ is recursive:

- $\ulcorner$(T super-is present)$^X\urcorner$ is defined as $\ulcorner$T $\underline{is}$ present at X$\urcorner$.

  (In general, if P is sentence with no tense operators or logical vocabu-lary then $\ulcorner(P)^X\urcorner$ is the result of making all verbs in P tenseless and then appending $\ulcorner$at X.$\urcorner$)

- $\ulcorner$(It is not the case that P)$^X\urcorner$ is $\ulcorner$It is not the case that (P)$^X\urcorner$.

- $\ulcorner$(P and Q)$^X\urcorner$ is $\ulcorner$(P)^X and (Q)$^X\urcorner$;   similarly   for $\ulcorner$(P or Q)$^X\urcorner$ and $\ulcorner$(If P then Q)$^X\urcorner$.

- $\ulcorner$(It super-was the case that P)$^X\urcorner$ is $\ulcorner$There is a supertime Y earlier than X such that (P)$^Y\urcorner$, and similarly for "super-will."

- $\ulcorner$(For all Y, P)$^X\urcorner$ is $\ulcorner$For all Y that $\underline{exist}$ at X, (P)$^X\urcorner$.

Before proceeding let me say something about the importance of these rules. The use of supertense operators in MST-Supertime is an adequate replacement for talk of supertime in MST-Supertime only if these translation rules exist. For those operators are adequate only if, roughly speaking, anything we want to say in MST-Supertime about how things vary from supertime to supertime can be said using the supertense operators. And we can say in a supertensed language every-thing we want to say by speaking of supertime only if there is a way to translate each language into the other.[6]

---

[6] The existence of the translation rules is necessary but not sufficient for the supertense opera-tors to be adequate. It had better also be true that a set of supertensed sentences is consistent in MST-Supertense if and only if its translation into the language of MST-Supertime is consistent. To satisfy this requirement MST-Supertense needs to say something about the logic of its supertense operators. A list of what axioms these operators should satisfy to meet these requirements may be found in chapter 3 of Burgess's *Philosophical Logic*. Burgess is discussing the logic of "ordinary" tense

Let me return to the main thread. So far I have just said that the passage of time in MST-Supertime amounts to the truth of the following:

(8) For all times M, if M super-is later than the present time, then it super-will be the case that (M super-is present).
(9) For all times M, if M super-is earlier than the present time, then it super-was the case that (M super-is present).

Let N be the time that is present at supertime S; then the translations of (8) and (9) into the language of MST-Supertime are

(8*) For all times M (that exist at S), if M is later than N (at S) then there is a supertime R later than S such that M is present at R.
(9*) For all times M (that exist at S), if M is earlier than N (at S) then there is a supertime R earlier than S such that M is present at R.

(Claims (8*) and (9*) are a little strange, for they speak of whether time M exists at supertime S, and is earlier than N at supertime S. But which times exist and which are later than which others are not the sorts of things that change as time passes in the moving spotlight theory. So time M exists, and is earlier than time N, at some supertime if and only if it exists and is earlier than N at all supertimes. For convenience, then, we could drop the qualifier "at S" from these clauses. That is why I have put them in parenthesis.)

Statements (8*) and (9*) are not adequate to capture the passage of time in MST-Supertime. For it is consistent with (8*) and (9*) that there be two future times O and P, with O earlier than P, such that P gets to be present before O does. Figure 7.1 illustrates this scenario (as before, an arrow points from a point in supertime to the time that is present at that point).

This is not the only problem with (8*) and (9*). They also fail to guarantee that each point in supertime has a time that is present relative to it. So they fail to guarantee that as time passes some time or other is always present.

What we want is supertensed sentences the translations of which are (collectively) consistent only with Figure 4.3 and those relevantly like it. Figure 4.3 is reproduced here as Figure 7.2; figures relevantly like it are those in which

operators, not supertense operators, but he is answering the analogous question, namely, what axioms must those tense operators satisfy to guarantee that, if time exists and the operators are given reductionist truth conditions, time is linear and dense and has no past or future endpoints? (I said that these axioms give the logic of the tense operators, but on some views these axioms should not be regarded as logical truths, since they correspond to substantive claims about the structure of time or supertime. At best they are necessary truths. If one takes this view then "S is consistent" in the previous paragraph should be understood to mean "it is metaphysically possible that all members of S be true" rather than "no contradiction follows logically from S.")

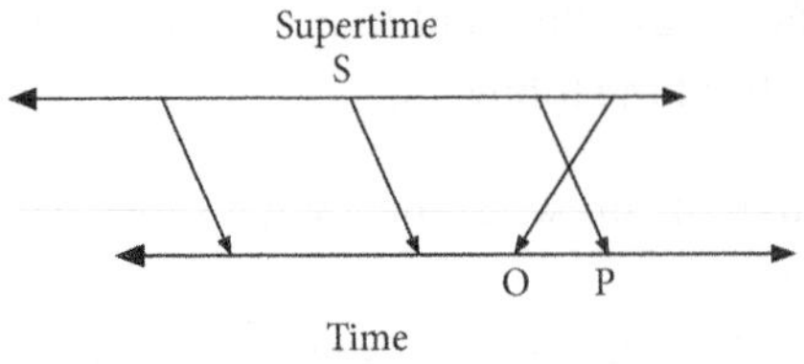

Supertime
S

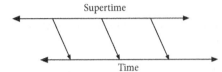

Figure 7.1 Incorrect passage.

O   P

Time

Supertime

Figure 7.2 Correct passage.

Time

each supertime has an arrow leaving it, each time has exactly one arrow "hit-ting" it, and arrows do not cross. These are sufficient for presentness to move continuously into the future.[7]

One set of sentences that does that trick uses nested supertense operators:

(10) Exactly one time super-is present; it super-is always going to be the case that exactly one time super-is present; it super-has always been the case that exactly one time is present.[8]

(11) T1 super-is later than the present time, and T2 super-is later than T1, if and only if it super-will be the case that (T1 super-is present and it super-will be the case that (T2 super-is present)).

(12) T1 super-is earlier than the present time, and T2 super-is earlier than T1, if and only if it super-was the case that (T1 super-is present and it super-was the case that (T2 super-is present)).

To see that (10) through (12) do what we want look at their translations:

(10*) Exactly one time is present at X, for each supertime point X.

(11*) T1 is later than N (at S), and T2 is later than T1 (at S), if and only if there is a supertime R later than S such that T1 is present at R, and there is a supertime Q later than R such that T2 is present at Q.

(12*) T1 is earlier than N (at S), and T2 is earlier than T1 (at S), if and only if there is a supertime R earlier than S such that T1 is present at R, and there is a supertime Q earlier than R such that T2 is present at Q.

---

[7] The conditions are equivalent to the claim that there is a bijective, strictly increasing function taking each supertime to the time that is present at it. Any such function is continuous.

[8] "It super-is always going to be the case that ..." is defined as "It is not the case that (it super-will be the case that (it is not the case that ...))." "It super-has always been the case that ..." is defined similarly.

What must a supertime diagram look like to make (10*) through (12*) true? By (10*) there must be one, and only one, arrow starting at each point in supertime, pointing to the time that is present at that point. By the left-to-right directions of (11*) and (12*) each time is "hit" by some arrow. By the right-to-left direction of (11*) and (12*) if supertime Q is earlier than supertime R then the time that is present at Q is earlier than the time that is present at R—arrows do not cross. I have a final side-note on (10) through (12). This statement of a version of the moving spotlight theory uses the predicate "X super-is earlier than Y." Some philosophers think that a better version of the moving spotlight theory will not treat this predicate as fundamental. A version that does not can define it as "Either Y super-is present and X super-was present, or it super-will be the case that (Y super-is present and X super-was present), or it super-was the case that (Y super-is present and X super-was present)."[9] A theory that does not treat "earlier than" as fundamental will need more complicated statements to collectively characterize the robust passage of time.

For various reasons we might want a version of the moving spotlight theory that does assign a rate to the passage of time. Can any version of MST-Supertense do that?

7.3 Seconds and Superseconds

One way to get a rate of passage into MST-Supertense is to expand the vocabulary available to state the theory. So far the theory has only non-metric tense operators. A seemingly more powerful theory also makes use of metric tense operators. With metric tense operators the theory can say not just that such-and-such super-will be the case, but can also say how long it super-will be until such-and-such is the case.

What is a metric tense operator? I want to approach this question slowly.

The tense operators in MST-Supertense replace the talk of supertime that appears in MST-Supertime. So let us start by seeing what that theory can say about the rate of time's passage. If time does pass at a definite rate it will be defined like this:

Let T_1 and T_2 be two times; let P_1 and P_2 be the points in supertime relative to which T_1 and T_2 are, respectively, present; then the average rate of time's passage from T_1 to T_2 is the temporal separation between T_1 and T_2 divided by the "supertemporal" distance between P_1 and P_2.

Suppose that we decide to measure temporal distances in seconds. We also need some unit for measuring supertemporal distances. Suppose we

[9] This idea goes back to McTaggart, *The Nature of Existence,* volume 2 (p. 271).

choose a unit and call it the "supersecond." Then the rate of time's passage will be some number of seconds per supersecond.

It is worth noting that this definition assumes that supertemporal separation, like temporal separation, is a measurable quantity. In fact the definitions of the units for measuring temporal and supertemporal separation—of the second and the supersecond—themselves presuppose that these are measurable quantities. To define the second we chose two particular times T and S and decreed that any two times shall be N seconds apart if and only if the temporal separation between them is N times as great as the temporal separation between T and S.[10] Obviously this makes sense only if, for any two times, there is some number such that the temporal separation between them is that number of times as great as the temporal separation between T and S. Similarly, the definition of the supersecond will name two points in supertime and say that any two points in supertime are N superseconds apart if and only if the supertemporal separation between them is N times as great as the supertemporal separation between the reference points.

MST-Supertime does not need to say that supertemporal separation is a measurable quantity. It could say instead that "the supertemporal separation between P and Q is N times as great as the supertemporal separation between X and Y" makes no sense. Then we would have a version of MST-Supertime in which time passes, but not at any definite rate.[11]

But let us set that version of MST-Supertime aside and work with the version that does say that supertemporal separation is a measurable quantity. A definition of the supersecond requires a choice of reference points. What reference points should we choose? One way to go is to choose any two times T and S that are one second apart, and then let the reference points be the points P and Q in supertime relative to which T and S are present. (Then by definition the temporal separation between P and Q will be one supersecond.) With this definition it follows that in MST-Supertime time always passes at a rate of one second per supersecond. But is worth emphasizing that any two points in supertime could have been chosen as the reference points. So the supersecond could have been defined to make the rate of time's passage be any number of seconds per superseconds one wanted.

---

[10] The current definition used by the international bureau of weights and measures does not actually name two particular instants. Instead the second is defined by reference to the frequency of the radiation emitted by cesium atoms under certain conditions—not some particular bit of radiation emitted on some particular occasion by some particular cesium atom, but that kind of radiation in general. This, of course, only makes for a good definition if the frequency of that kind of radiation is always the same. The "certain conditions" the bureau describes ensure that this is so.

[11] Are we so sure the temporal separation is a measurable quantity? Our most successful physical theories assume this, but some have attempted to formulate alternatives that do not. Barbour presents a non-technical defense of these alternatives in *The End of Time*.

The number used to measure the rate of time's passage is a matter of convention. The substantive claim is that time passes at a constant rate.

Now in MST-Supertense we eliminate talk of supertime and replace it with supertense operators. How can we capture, with these operators, the claim that time passes at one second per supersecond? One natural idea is to use metric supertense operators like these:

It super-will be the case in r superseconds that . . .

It super-was the case r superseconds ago that . . .

There is one of each of these for each positive real number r. So this theory has infinitely many tense operators in its fundamental language. That is a lot of basic pieces of vocabulary, but the sheer number of them is not the real problem.

I will get to the real problem shortly, but first let us see what MST-Supertense can now say about the rate of time's passage. To claims (10) through (12) above we can add the claim

(13) Let N be the present time and M any time to the future of N. If M is r seconds to the future of N, then it super-will be the case in r superseconds that M is present. (Similarly for past times.)

Assuming that metric tense operators have the logical properties one expects,[12] (13) entails that time passes at one second per supersecond.

I do not like this theory, but not for any of the reasons that Broad or Smart advanced. Broad, recall, claimed

(3) "How fast does presentness move along the series of times?" (which just means "How great a time-lapse will presentness have traversed in unit time-lapse?") is a meaningless question.

But in MST-Supertense

How fast does presentness move?

does not mean

How great a time-lapse will presentness have traversed in unit time-lapse?

Instead it means

---

[12] For example, it super-will be the case in $r + s$ superseconds that X if and only if (it super-will be the case in r superseconds that (it super-will be the case in s superseconds that X)).

How great a time-lapse will presentness have traversed in unit supertime-lapse?

So the question that (1) and (2) say must be meaningful, if the moving spotlight theory is true, is not the question that (3) says is meaningless. Furthermore I do not see any reason to regard the question

How great a time-lapse will presentness have traversed in unit supertime-lapse?

as meaningless.

What about the second argument, the one that came from things Smart and Schlesinger said? Two of its premises were

(5) If the moving spotlight theory is true then there must be something informative to say about the rate of time's passage. A claim about the rate of time's passage that is true "by definition" cannot be the correct answer to this question (as opponents of the moving spotlight theory intend it to be understood).

(7) If the moving spotlight theory is true then "one second per second" is the only potential answer to "How fast does time pass?" If the theory cannot give that answer it can give no answer.

Statement (7) is clearly false in MST-Supertense. What is more, the claim the theory makes about the rate of time's passage—claim (13) above—is substantive. It is not true by definition. A perfectly consistent alternative to MST-Supertense might say that while the time five seconds later than the present super-will be present in five superseconds, the time seven seconds later super-will be present in just six superseconds. Time will speed up! I cannot imagine that a defender of the moving spotlight theory will want to accept a theory that says that the rate of time's passage changes, but he is not precluded from accepting such a theory by the meanings of his words.

However, the features that MST-Supertense must have in order to evade Broad's and Smart's arguments give the theory new defects. Surely the tense operators "It super-will be the case in r superseconds" are not fundamental pieces of vocabulary for describing reality! Names for units of measurement, like "meter," "second," and "supersecond," do not belong in the fundamental language. Nature does not single out a special unit of measurement.[13] Now it

---

[13] More cautiously, nature does not single out a special unit for measuring time or supertime, and does not single out a special unit for measuring distances in Euclidean geometry. There is some reason to think that there is a special unit for length in hyperbolic geometry.

is easy to get "second" out of the fundamental vocabulary. We have already seen how to do it. Let the fundamental vocabulary for talking about temporal separation be the two predicates "the temporal separation between T and S is the same as the temporal separation between U and V" and "S is temporally between R and T." From these can be defined the infinite family of predicates "the temporal separation between T and S is r times as great as the temporal separation between U and V." I showed earlier how to use these to define "the temporal separation between R and S is r seconds."[14] But this method for getting "second" out of the fundamental vocabulary works only if there are such things as times. Since MST-Supertense explicitly says that supertime does not exist, the analogous technique for defining "supersecond" in more fundamental terms does not work. (Although I have written off MST-Supertime this is a place where it has the advantage. Talk of superseconds can be eliminated in MST-Supertime in exactly the same way we eliminate talk of seconds.)

There are other techniques one could use to try to rid MST-Supertense's fundamental vocabulary of "supersecond." One idea is to try to define the metric tense operators using the non-metric tense operators.[15] Here is one way to do this:

> It super-will be the case in r superseconds that X $=_{df}$ It super-will be the case that (the time that super-is present is r seconds later than the time that super-is now present, and X).

The references to seconds can then be eliminated from the fundamental language in the way I indicated above.

The only problem I see with this theory is that it conflicts with the premise

(5) If the moving spotlight theory is true then there must be something informative to say about the rate of time's passage. A claim about the rate of time's passage that is true "by definition" cannot be the correct answer to this question (as opponents of the moving spotlight theory intend it to be understood).

For on this new version of MST-Supertense "time passes at one second per supersecond" is true by definition.

---

[14] The philosopher who did the most to emphasize the philosophical interest of this method for eliminating talk of seconds (and of numbers)—a method that had been around for a long time in mathematics—was Hartry Field, especially in his book *Science Without Numbers*.

[15] Another idea is to take neither "It super-will be the case that in r superseconds ..." nor "It super-will be the case that ..." as fundamental, but some other, stranger tense operators as fundamental. Sider makes a suggestion along these lines in a footnote (*Writing the Book of the World*, p. 240). He says that he does not know how well that suggestion works; I will not explore it here.

Now I do not think this shows the theory to be false. I think it is reasonable for a believer in objective becoming to reject (5). Still, it would be nice if a believer in objective becoming were not forced to say either that time passes but not at any particular rate, or that it is a definitional truth that time passes at one second per supersecond.

There is an alternative. So far I have been talking about how well these arguments work against MST-Supertense. But there is also MST-Time. That theory does not need to have any truck with the supersecond or with metric tense operators. In it the question

How fast does presentness move?

means just what Broad wanted it to mean, namely

How great a time-lapse will presentness have traversed in unit time-lapse?

Here is a slow and careful statement of the definition of the rate of time's passage in MST-Time:

> Let T and S be times that are one second apart. Then the average rate at which time passes between T and S, in seconds per second, is equal to number of seconds between the time that is present from the perspective of T and the time that is present from the perspective of S.

Since from the perspective of any time that time is present, the result is that time passes at one second per second.

The fact that MST-Time does without talk of the supersecond might seem to be a mixed blessing. MST-Supertense's use of "supersecond" had some benefits. It allowed the theory to side-step the worries Smart raised about the intelligibility, or informativeness, of saying that time passes at one second per second. MST-Time must face those worries squarely.

7.4 Arguments from Dimensional Analysis

Recently several philosophers have tried to make those worries more precise. Peter van Inwagen articulates them forcefully in the following passage:

If time is moving (or if the present is moving, or if we are moving in time) how *fast* is whatever it is that is moving moving? No answer to this question is possible. "Sixty seconds per minute" is not an answer to this question, for sixty seconds *is* one minute, and—if x is not 0—x/x is always equal to 1 (and 'per' is simply a special way of writing a division sign). And '1' is not, and cannot ever be, an answer to a question of the form, "How fast is such-and-such moving?"—no matter what 'such-and-such' may be. "One meter per second" is an answer, right or wrong, to the question, "How fast is the cart moving?" . . . ; 'One', 'one' "all by itself," 'one' *period*, 'one' *full stop*, can be an answer only to a question that asks for a number; typically, these will be questions that start "How many . . .". 'One' can be an answer, right or wrong, to the questions "How many children had Lady Macbeth?", "How many Gods are there?", and "How many minutes do sixty seconds make?"; 'one' can never be an answer, not even a wrong one, to *any* other sort of question—including those questions that ask 'how fast?' or 'at what rate?' Therefore, if time is moving, it is not moving at any rate or speed. And isn't it essential to the idea of motion that anything moving be moving *at some speed* . . .? (*Metaphysics*, p. 75)

I call the main argument van Inwagen presents here The First Argument from Dimensional Analysis.[16] (There will be a second. I will explain the name presently.) The argument may be formalized like this:

(1) One second per second is (identical to) the result of dividing one second by one second.
(2) The result of dividing one second by one second is the number 1.
(3) The number 1 is not a rate.
(C) Therefore, nothing (the passage of time or anything else) can happen at a rate of one second per second.

To get from here to the conclusion that there is no such thing as the passage of time van Inwagen needs additional premises. These are the ones he seems to use:

(4) If time passes it passes at some rate.
(5) If time passes at some rate then that rate is some number of seconds per second.

I have already said that the moving spotlight theorist may reject (4) and (5). But I have also said that he may have a better theory if he can accept them.

Even if (4) and (5) are true the First Argument from Dimensional Analysis fails. Premise (1) is false. It rests on a false view about the nature of measurable quantities and about how numbers are used to measure them. But explaining why that view is false will take some setting-up.

A *measurable quantity* (quantity for short) is a determinable property whose family of determinates has a certain kind of structure.[17] What kind of structure makes a property a quantity? It is easiest to explain with an example. Length is a determinable property; the various maximally specific lengths, like the length property shared by all and only things that are precisely as long as the

---

[16] Another recent presentation and defense of the argument is in Olson, "The Rate of Time's Passage."

[17] Again, I am a nominalist. But here, as I have done earlier, I am going to speak as if there are such things as properties. It makes it easier to say what I want to say.

standard meter, are its determinates. I will call these determinate properties *values* of length. The length values have a "ratio structure." What does that mean? Suppose, for example, that I have two sticks and the first one is twice as long as the second one. This relation between the sticks is reflected in the ratio of the length values the sticks instantiate. The ratio of the length value had by the first stick to the length value had by the second is 2. What is true here of length is true of any quantity:

> A determinable property is a quantity if and only if for any two values of that determinable there is a number that is the ratio of the first value to the second, and these ratios structure the values so that their structure is relevantly like the structure of the positive real numbers.[18]

To say anything useful about the values of a quantity we need an informative way to refer to them. (I can always refer to my current mass as "the mass value I currently instantiate," but this is of no practical use.) It would be especially useful if we assigned names systematically so that relations among names of values reflected the relations among the values themselves. The best way to do this is to use numbers as names for quantity values. A *scale* for measuring a quantity is an assignment of numbers to values of that quantity. A *faithful scale* (I will only be interested in faithful scales) is one with the following feature: if n names value x and m names value y then n/m is equal to the ratio of x to y. That is, dividing the number that names the first value by the number that names the second gives the ratio of the first value to the second. For convenience we can use "x/y" to name the number that is the ratio of x to y; then the numbers n, m assigned to values x, y on a faithful scale satisfy $n/m = x/y$. But we should be aware that "/" has different (though clearly related) meanings on the two sides of this equation. On the left side it stands for an operation on numbers while on the right it stands for an operation on quantity values.

To set up a faithful scale for measuring some quantity we only need to select a unit. A unit is just a particular value for that quantity, the value that shall be

---

[18] "Relevantly like" here means that the set of values is isomorphic to the additive semi-group of the positive real numbers. (Among the real numbers the number 1 is special, but its specialness comes from its role in the multiplicative structure of the reals. Since the structure of a set of quantity values is like the additive structure of the (positive) reals, there is no quantity value that plays the special role in the set of values that the number 1 plays in the reals.)

The general definition of quantity I have given really applies only to positive scalar quantities. I will not say anything about vector quantities or quantities with non-positive values in this chapter.

A broader notion of (scalar) quantity counts determinables without ratio structures as quantities. Perhaps the values just need to be ordered, the way hardness properties are. (Stevens describes several structures one might want to allow quantities to have in "On the Theory of Scales of Measurement.") But the narrow notion of quantity is all we need for the purposes of this section.

assigned the number 1 on the scale. The ratios between other values and the unit then determine the numbers assigned to all other values. To set up a scale for length, for example, we might choose a particular bar made of a platinum-iridium alloy. This bar has some value L for length. We say that the number assigned to any length value v on our scale is equal to v/L.[19] To make clear what scale we are using when we attribute numerical lengths to things we name this unit "the meter" and report lengths as "n meters."

In many contexts (scientific research, physics homework) we need numerical values for lots of different quantities.[20] So we need scales for measuring all of those quantities. We could go through the quantities one by one, setting up un-related scales for measuring each one. But that would be a lot of work and would make for a lot of extra computation when solving problems. Suppose, for exam-ple, that we measure lengths in meters and durations in seconds. We also need a scale for measuring speed. Suppose we choose some arbitrary unit for measuring speed—say, the speed of sound through air under standard conditions. We name this unit "the bleg" and report speeds on this scale as "n blegs." Then if we want to measure something's speed it is not enough to measure how many meters it trav-els in one second. We would also need to figure out how many meters something moving at 1 bleg travels in a second, and then do some calculations. What a pain.

It is easier to set up a system of scales (also called a system of units). In a system of scales the units for the scales are systematically related. To set up a system of scales we choose a select few quantities and set up scales for measuring those quantities. The units we choose for these scales are the fundamental units of the system. These fundamental units then determine derived units for the scales of other quantities. The derived unit for a quantity is a value of that quantity that is singled out by the values used as fundamental units. It is easiest to explain this by looking at an example. Suppose that our system uses the meter

---

[19] Given certain assumptions, this way of defining a scale is equivalent to the way I described in Section 7.3. There the example was the definition of a scale for measuring temporal separation; the analogous definition of the meter scale for length says that "x and y are r meters apart" means "x and y are r times as far apart as p and q," where p and q are at the endpoints of the standard meter.

If the "certain assumptions" fail then these ways of defining a scale are not equivalent. One of those assumptions, of course, is that there are such things as properties. Another is that there are "enough" things that have values of the quantity being defined. For example, the second way of defining a scale for length uses predicates like "x and y are 2 times as far apart as p and q." This predicate is standardly defined to mean "there is a point z between p and q such that x and y are as far apart as p and z, and also as far apart as z and q." If there are no points of space between p and q then there is trouble. (The trouble is more likely to arise when defining scales for mass or temperature than when defining scales for length or duration.) These issues are important to the debate over whether there really are such things as properties, but they are not important here.

[20] Much of what follows here is adapted from chapter 1 of Barenblatt, *Scaling, Self-Similarity, and Intermediate Asymptotics*.

as the fundamental unit for length and the second as the fundamental unit of duration. The derived unit for speed is then the speed value had by something that moves a distance of 1 meter during 1 second (and does so without changing its speed during that second). The scale we end up with is the meters per second scale for speed, and we report speeds as "n meters per second" (often abbreviated as "n m/s"). This scale for speed is obviously superior to the bleg scale. When we use the meters per second scale, once we have measured how many meters something travels in one second we automatically know its speed in meters per second. Clearly a system of scales saves an enormous amount of time and effort. If we choose fundamental units for length, duration, and mass we get derived units for lots of quantities (speed, acceleration, frequency, force, ...), and can measure something's value for one of these quantities just by measuring lengths, durations, and masses.

Why, when we use this system of scales, do we report speeds as "n meters per second"? Because reporting them this way indicates the dimension of speed. But what is meant by talk of the dimension of a quantity? Answering this question requires some more definitions. Suppose we have decided to adopt a system of scales that measures lengths, durations, and masses in fundamental units—we think that, for our purposes, these are the quantities it is most convenient to have fundamental units for. (In other contexts it may be more convenient to choose different quantities to measure in fundamental units.) There are lots of systems of scales we could adopt that all measure length, duration, and mass in fundamental units. Let a class of systems of scales be a collection of systems each of which chooses the same quantities to measure in fundamental units. A system that uses the meter, second, and kilogram as fundamental units is in the same class as a system that uses the foot, the hour, and the gram as fundamental units. Systems like these that choose length, duration, and mass as the quantities that shall be measured in fundamental units are in the "length-duration-mass" class.

Not all systems of units are in the length-duration-mass class. One reason to be interested in other classes is that the systems in the length-duration-mass class are not adequate to measure quantities like electric charge or electric current. But even if we ignore quantities that cannot be measured by systems in the length-duration-mass class, there are other classes of systems that might be of interest. The systems of units in these classes choose different quantities to measure in fundamental units but still measure all of the quantities that can be measured by systems in the length-duration-mass class. An example is the length-duration-force class. A system of units in this class chooses a fundamental unit to measure force. Its unit for mass is a derived unit. It is the value of mass with the property that when unit force is applied to a body with that mass it produces a (constant) unit acceleration.

I myself tend to think that some select set of quantities is "metaphysically fundamental," and that all other quantities are "derived from" the fundamental quantities. If this view is correct then a class of systems the members of which measure only basic quantities with fundamental units better reflects the structure of reality than other classes. But nothing in this section turns on whether this view is true.

We might, on some occasion, decide to switch from one system in the length-duration-mass class to another. Maybe we decide to switch to a system that measures lengths in centimeters instead of meters. Every length value will be assigned a number in the new system that is different from the one it was assigned in the old. But it is easy to calculate what these new numbers are. There are 100 centimeters in a meter, so any length that is assigned a number n on the meter scale is assigned the number $100n$ on the centimeter scale. But it is not just length values that are assigned different numbers in the new system; lots of other quantities, quantities measured in derived units, are too. The numbers assigned to areas, speeds, forces, and so on are all different. But the way that these numbers change is completely determined by the way the numbers assigned to the values of the quantities measured in fundamental units change. The *dimension function* of a quantity contains complete information about these changes. (Strictly speaking, a quantity does not have a dimension function *simpliciter*; it has a distinct dimension function for each class of systems of units. When I speak of "Q's dimension function" without qualification I shall just mean Q's dimension function relative to whichever class of systems is contextually salient.)

Here is how the dimension function works. I will follow common practice and use "$[Q]$" as a function-symbol standing for the dimension function of quantity Q. The dimension function of a quantity is a function from n-tuples of numbers to a single number. Suppose we change from one system of units in the length-duration-mass class to another. Then there is a number that is the ratio of the unit for length in the old system to the unit for length in the new system. (When we switch from meters to centimeters this ratio is 100.) Let L be this ratio. Similarly, let T be the ratio of the old unit of duration to the new and M the ratio of the old unit of mass to the new. Then the dimension function for quantity Q takes these three numbers as inputs. As output it gives the factor by which the numbers assigned to values of Q change when we move from the old system of scales to the new. So the dimension function satisfies this equation ("$\#v$" abbreviates "the number assigned to value v of quantity Q"):

$$\left(\#v \text{ in the new system}\right) = [Q](L, T, M) \times \left(\#v \text{ in the old system}\right). \qquad (7.1)$$

For example, the dimension function for speed operates on the triple of numbers (L, M, T) and gives back the number L/T. (We see here how reporting velocities as "n m/s" indicates what the dimension function of speed is.) For an example of this dimension function in action, suppose that we change from the MKS system to the CGS system. We change from using the meter, the kilogram, and the second as fundamental units to using the centimeter, the gram, and the second as fundamental units. The ratio of the meter to the centimeter is 100; the ratio of the kilogram to the gram is 1000; and the ratio of the second to itself is 1. So $L = 100, M = 1000$, and $T = 1$, so $[\text{speed}](L, M, T) = [\text{speed}](100, 1000, 1) = 100/1 = 100$. This means that if a speed value is assigned the number N in the MSK system it is assigned the number 100N in the CGS system. Writing it out in one line, the speed value assigned 5 in the MKS system is assigned

$$[\text{speed}](100, 1000, 1) \times 5 = 100/1 \times 5 = 500$$

in the CGS system. That is, 5 meters per second is identical to 500 centimeters per second.

Relative to any given class of systems of units some quantities are dimensionless. Q is dimensionless if and only if $[Q] = 1$, that is, if and only if the numbers assigned to values of Q by each scale in the system are the same. There are lots of examples of quantities that are dimensionless in the length-duration-mass class. When a weight is hung from the lower end of a bar that is oriented vertically and fixed at the upper end the bar lengthens. Strain is a dimensionless quantity had by the bar. The bar's strain is determined by how much it lengthens. If a bar's strain is .01 (if its value for strain is assigned this number by any system of units), for example, and its value for length when no weight is attached is assigned the number r on some scale, then its length when the weight is attached is assigned the number 1.01r. The Reynolds number of a fluid flowing over a surface is another example of a dimensionless quantity. The Reynolds number characterizes the relative importance of viscosity in the flow, and is determined by the fluid's velocity, density, viscosity, and the size of the surface. (Roughly speaking, when a fluid's value for the Reynolds number is assigned a small number then viscosity is important.)

Though I sometimes omit the relativization, it is worth emphasizing again that quantities are only dimensionless or dimensionful relative to a class of systems of units. Strain is dimensionless in the length-duration-mass class, but we could adopt a system from the class of systems that measure length, duration, mass, and strain in fundamental units. Strain is not dimensionless in this system. It would, of course, be annoying to use a system like this.[21]

---

[21] However, given the choice between a system of units in which a given quantity is dimensionless and a system in which that quantity has dimensions, the scientific community does not always

(If some quantities are fundamental then one can make sense of the idea that some quantities are dimensionless *simpliciter*. A quantity is then dimensionless *simpliciter* if and only if it is dimensionless relative to a class that measures only fundamental quantities with fundamental units. But, again, the notion of fundamentality does not play a role in the argument of this section.)

Strain and the Reynolds number are just two of the many dimensionless quantities scientists are interested in. The existence of multiple dimensionless quantities shows that distinct quantities can have the same dimension. There are also distinct dimensionful quantities with the same dimension. Thermal diffusivity, for example, is a quantity that influences how heat flows through a substance. It is measured in square meters per second. The kinematic viscosity of a fluid is also measured in square meters per second. Among other things, the kinematic viscosity controls how momentum "diffuses" through the fluid.[22] Kinematic viscosity is a property only fluids can have, while both solids and fluids have thermal diffusivities; and the kinematic viscosity of a fluid need not be assigned the same number as its thermal diffusivity by any system of scales. Clearly thermal diffusivity and kinematic viscosity are different quantities.

Since distinct quantities can have the same dimension, expressions like "4 square meters per second" are not always used to denote the same thing. In one context this expression may name a value of kinematic viscosity, in another it may name a value of thermal diffusivity.

Now we are in a position to see why the First Argument from Dimensional Analysis fails. Here again is the argument:

(1) One second per second is (identical to) the result of dividing one second by one second.
(2) The result of dividing one second by one second is the number 1.
(3) The number 1 is not a rate.
(C) Therefore, nothing (the passage of time or anything else) can happen at a rate of one second per second.

---

choose the first system. The main example of this is entropy. The SI system contains a fundamental unit for measuring temperature. As a consequence, in the class to which the SI system belongs the dimension function of entropy is $[E]/\Theta$, where $[E]$ is the dimension function for energy and Θ is the ratio of units for temperature. But entropy is dimensionless in the class of systems that is just like the class to which the SI system belongs except that it does not measure temperature in fundamental units. (The dimension function for temperature in this class is the same as the dimension function for energy.)

[22] Talk of momentum diffusing through a fluid is common in certain parts of physics. But it cannot be literally true. Only stuffs can literally diffuse, but momentum is a property, not a stuff. Formally momentum can be represented by an assignment of numbers to points of space (giving the "density of momentum" at each point), and these assignments obey an equation formally identical to a diffusion equation.

The first challenge we face when evaluating this argument is that it contains the description "the result of dividing one second by one second." The primary use of the word "division" has it naming an operation on numbers, not an operation on quantity-values. Still, perhaps we can grant that there is an extended use of "division" so that "the result of dividing v by u," where v and u are values of a single quantity, denotes the ratio of v to u.

Having granted this, premise (2) is true. And (3) is certainly true. Rates are values of certain special quantities, not numbers. But what about (1)? Given what I have said about the meaning of expressions of the form "N Xs per Y" this premise is false. For in MST-Time "one second per second" is used to name a value of the quantity speed-through-time, just as "one meter per second" is used by everyone to name a value of the quantity speed-through-space. So (1) says that a value of the quantity speed-through-time is identical to 1 second divided by 1 second. Given (2) this amounts to saying that a value for a certain quantity is identical to a number. But while numbers are assigned to values of quantities by scales of measurement, no number is identical to a value of a quantity.

Friends of the First Argument from Dimensional Analysis may reply that I am begging the question. I have just assumed that "one second per second" is used, in MST-Time, to name the value of a quantity, but the whole point of their argument is that it cannot be used this way. So let us back up and see what defense of (1) they offer.

Van Inwagen, in the quoted passage above, says that "per" as it occurs in expressions like "10 meters per second" is "simply a special way of writing the division sign." If that were true then (1) would be true. His view seems to be that there is a single function, division, that can take as an argument either a pair of numbers, or a pair of values of (possibly distinct) quantities, or a pair of a value of a quantity and a number (this case will not matter). When you take two numbers and divide them you get another number. When you take values of the same quantity and divide them you get a number. When you take values of different quantities and divide them you get a value for some other quantity. Divide one second by one second and you get a number. Divide one meter by one second and you get a speed.

These claims are false. For if they were true then there could not be distinct quantities with the same dimension.[23] These claims entail, for example, that dividing the square meter by the second yields a value of some single quantity. But there are many quantities the values of which can be named by "n square meters

---

[23] Similar points are made in Barenblatt, *Scaling, Self-Similarity, and Intermediate Asymptotics* (p. 33), and Palacios, *Dimensional Analysis* (p. xiii).

per second." (Thermal diffusivity and kinematic viscosity are the two examples I gave earlier.) And, in general, for any dimension you choose, there are plenty of distinct quantities that have that dimension.

The correct way to understand "*n* meters per second" is the way I described in the last section. "Meters per second" indicates what system of units is being used to measure speed, and indicates what the dimension function of speed is.

Computations of the following sort might be thought to support the view I oppose:

$$1\frac{\text{meter}}{\text{sec}} \times 60\frac{\text{sec}}{\text{min}} = 60\frac{\text{meters}}{\text{min}}. \tag{7.2}$$

It looks here like we are first dividing meters by seconds and then dividing seconds by minutes, and that the product of the result is some number of meters divided by a minute. But there is no problem understanding these computations on my view. We have seen them before. The dimension function of speed is here being used to find the number assigned to the speed value 1 meter per second in the system of units that uses the minute as its unit for time. So equation (7.2) is just an instance of equation (7.1) (mirror-reversed around the identity sign and with the numeral "1" omitted in several places), where labels have been attached to indicate which number goes with which system of units. On my view, made fully explicit equation (7.2) looks like this:

$$1\frac{\text{meter}}{\text{sec}} \times [\text{speed}]\left(1, 1, \frac{60 \text{ sec}}{1 \text{ min}}\right) = 60\frac{\text{meters}}{\text{min}}.$$

Turning our attention back to the original equation (7.2), Sometimes people "cancel units" when performing the computation this equation represents:

$$1\frac{\text{meter}}{\cancel{\text{sec}}} \times 60\frac{\cancel{\text{sec}}}{\text{min}} = 60\frac{\text{meters}}{\text{min}}. \tag{7.3}$$

Here there is a stronger suggestion that the first expression is the result of dividing the meter by the second. Otherwise, how can you cancel the two occurrences of "sec"? Isn't (7.3) just like

$$\frac{1 \cancel{4}}{\cancel{4} \ 3} = \frac{1}{3}, \tag{7.4}$$

with the second taking the place of four, the meter taking the place of one, and so on? The answer is no. The analogy between (7.3) and (7.4) is merely formal. We should not take computations like the one in (7.3) seriously when doing metaphysics. Compare, for example, the following computation involving derivatives:

$$\frac{dy}{dt} = \frac{dy}{dx}\frac{dx}{dt}. \tag{7.5}$$

Equation (7.5) also looks like (7.4) (without the explicit cancelations). It looks like the left-hand side of (7.5), the derivative of y with respect to t, was obtained from the right-hand side by canceling dx. And this suggests that dy/dx is the result of dividing something named "dy" by something named "dx." But while early mathematicians may have believed that this is what is going on, in modern developments of calculus "dx" and "dy" do not name anything, and so dy/dx is not the result of dividing dy by dx. Some math books even refuse to use this notation so that students do not become confused. Still, this notation is computationally convenient. In certain contexts if we pretend that dy/dx is a division problem and manipulate it formally like a quotient then we will get the right answer.[24] This is also true, I claim, about "meters/sec."

Speed's dimension function is L/T (in the class of systems of units we normally use). But the fact that speed-through-time is measured in seconds per second indicates that this quantity's dimension function is $T/T = 1$—speed-through-time is dimensionless. We can report something's speed through time by saying that its speed through time is 1, or 17, or whatever. We need not mention the units that indicate its dimension.

The fact that speed-through-time is a dimensionless quantity is clearly lurking in the background of the First Argument from Dimensional Analysis. One can read (1) and (2) as an attempt to establish that speed-through-time is dimensionless. But reading them this way does not make the argument any better. It still mistakenly identifies the value of a (dimensionless) quantity with a number. It is better to start over and construct an argument against objective becoming that works directly with the notion of a dimensionless quantity. Huw Price seems to be getting at an argument like that when he discusses the rate of time's passage:

perhaps the strongest reason for denying the objectivity of the present is that it is so difficult to make sense of the notion of an objective flow or passage of time. Why? Well, the stock objection is that if it made sense to say that time flows then it would make sense to ask how fast it flows, which doesn't seem to be a sensible question. Some people reply that time flows at one second per second, but even if we could live with the lack of other

---

[24] In his book *Calculus on Manifolds* Michael Spivak complains about the confusions that this notation can cause (pp. 44–5). On the other side, one mathematician writes, "if one is concerned with computations and applications rather than the abstract theory, then one is quickly led to the conclusion that the Leibniz notation is incomparably more efficient" (Edwards, *Advanced Calculus*, p. 459). Of course, there are mathematical theories in which "dx" does have an independent meaning, namely the theory of differential forms and some theories of infinitessimals. This does not affect my point, which is that in "standard calculus" (7.5) is correct but does not involve canceling the dx's.

possibilities, this answer misses the more basic aspect of the objection. A rate of seconds per second is not a rate at all in physical terms. It is a dimensionless quantity, rather than a rate of any sort. (*Time's Arrow and Archimedes' Point*, p. 13.)

The crucial claim that Price makes is that no rate is a dimensionless quantity. It is the important premise in the Second Argument from Dimensional Analysis:

(6) One second per second is a value of a dimensionless quantity.

(7) No rate is a dimensionless quantity.

(C) Therefore, nothing (the passage of time or anything else) can happen at a rate of one second per second.

This argument (of course) goes back to C. D. Broad:

The numerical value of this ratio [of the distance presentness has traveled in time to the time it took to make the trip] is of no importance; it could always be given the value 1/1 by a suitable choice of the units in which we measure time-lapses in the two dimensions. The important point is that, whatever may be the numerical value, the ratio cannot possibly represent a rate of change unless its denominator measures a *lapse of time* and its numerator measures something *other than* a lapse of time in the same time-dimension. (*Examination of McTaggart's Philosophy*, volume 2, part 1, p. 278)

If the numerator and the denominator both measure lapses of time then the quantity measured is dimensionless. Like Price, Broad endorses (7).

The Second Argument is better than the First because it does not rest on any confusions about the meaning of expressions of the form "N Xs per Y." So what should we think of it? I am going to argue that we should reject (7), but first let us look at (6). Premise (6) is true—in the class of systems of units we normally use. Supposing that there is such a quantity as speed-through-time (this appears to be a presupposition of the argument), we could choose a system of units that measures this quantity with a fundamental unit. In that system it is not dimensionless and (6) is false. So (6) is like "Boston is on the left"—incomplete without some implicit or explicit relativization. One way to deal with this is to suppose that the premises are all implicitly relativized to some class of systems of units. But then it is hard to know what to make of (7). We are now reading (7) so it says "No quantity that is dimensionless in such-and-such a class is a rate." Why think that being a rate correlates with being dimensionless in some particular class of systems of units? Perhaps those who defend this argument think that some quantities are fundamental, and so think that some quantities are dimensionless *simpliciter*; then the premises do not need to be relativized, and this problem with (7) does not arise. But instead of pursuing this issue further I am going to set it aside and present grounds for rejecting (7) that are good whether or not some quantities are dimensionless *simpliciter*.

But first, what has been said in favor of (7)? Van Inwagen also discusses the Second Argument, and when he does he presents it in a way that makes (7) look tautologous. He writes the premise as "a speed or rate can never be a dimensionless number" (*Metaphysics*, p. 90). And of course numbers are not rates. But this is confused. The term "dimensionless number" is a category mistake. Only quantities can be dimensionless or dimensionful. Using the term "dimensionless number" encourages us to confuse values of dimensionless quantities with numbers. But values of dimensionless quantities are, again, not numbers.[25] (If they were, then there could not be distinct dimensionless quantities.[26]) When we keep this distinction clear, (7) does not look obvious.

To evaluate (7) we need a definition of "rate." This should not be controversial. For any quantity Q there is another quantity that is the rate of change of Q.[27] If we have a unit for measuring Q and a unit for measuring duration then the unit for measuring the rate of change of Q is measured in units of Q per unit of duration. If we write "$\dot{Q}$" for the rate of change of Q, then the dimension function for $\dot{Q}$ (in the length-duration-mass class) satisfies

$$[\dot{Q}](L, M, T) = \frac{[Q](L, M, T)}{T}.$$

It may tempting to think that a function of this form can never be the constant function always equal to 1; but this cannot be tempting for long. For if Q is any quantity that is measured in units of duration then its rate of change will be dimensionless. And there are plenty of uncontroversial examples of quantities like this. One example is the period of a pendulum, the time it takes for the bob to swing back and forth and return to its starting point. A pendulum's period is some number of seconds. Over time the length of the pendulum may change (due to, say, changes in the temperature), causing a change in the period. If the pendulum is part of a clock then the owner may be very interested in the rate at which its period changes; one can easily imagine a homework question in a physics class

---

[25] Van Inwagen also says (in the same footnote on p. 90) that the Second Argument is the same argument as the First, just put "much more neatly." This is wrong. The premise that numbers cannot be rates is distinct from the premise that dimensionless quantities cannot be rates. Van Inwagen identifies them because he does not distinguish between numbers and dimensionless quantities.

[26] In his discussion of this argument in "On the Passing of Time" Tim Maudlin also warns against this confusion by emphasizing that distinct quantities can have the same dimension (p. 117).

[27] Not all quantities that are rates are rates of change of some other quantity. A particle's speed is the rate at which its position is changing, but position is not a quantity as I have defined it. (Distances between positions are values of the quantity length (they are lengths of the lines between those positions), and it is these that figure in the definition of speed.)

asking you to calculate this rate. The rate can be reported as n seconds per second (and here n need not be 1). It can also be reported just as n—provided we remember that n here names a value of a quantity (the same value no matter what unit we use to measure duration), not a number.

To wrap up my discussion of the arguments from dimensional analysis let me step back from the technical details. Part of what makes the arguments seductive, I think, is the idea that one cannot figure out the rate at which objective becoming occurs merely by observing that a second is one second long. I think that the claims that one second per second is just a number, or at least is dimensionless, are attempts to prop this idea up. But MST-Time agrees with this idea. If you ask it to instruct you how to figure out the rate of time's passage the theory does not instruct you to find two times T and S that are one second apart, and then just look again to find out how many seconds apart they are. It asks you to do something more complicated. It asks you to find out how far apart are the times that are present from the perspective of T and from the perspective of S. And that these times turn out to be T and S themselves is not some sort of definitional truth.

7.5 Odds and Ends

A defender of the block universe can say that there is "anemic" passage of time. If any sense can be made of the question how fast this anemic passage occurs it is bound to have an uninformative answer. But that is okay, because there is no demand that in the block universe there be anything informative to say about the rate of time's passage.

I argued in Section 3.3 that the passage of time in presentism is not robust enough to be interesting. Further evidence for this claim is the fact that Prior, the ur-presentist, gave a deflationist answer to "How fast does time pass?" In one place he identifies this question with the question "How fast do I get older?" and says that the "obvious" answer is one year per year, one time-unit per time-unit ("Time After Time," p. 244). This (perfectly fine) answer to the question "How fast do I get older?" is an answer that a believer in the block universe could also give. But a believer in the moving spotlight theory (and, I think, in objective becoming more generally) should deny that the only thing "How fast does time pass?" can mean is this question about the rate at which one ages. They should want it to also have a more robust meaning.

In "How Fast Does Time Pass?" Markosian presents another answer the believer in objective becoming might give. He begins by suggesting that "talk about any rate essentially involves a comparison between two different changes." But, he continues, there is no restriction on what these changes are. So

whenever one gives the rate of some normal change in what is admittedly the standard way, i.e., in terms of the pure passage of time, then one has likewise given the rate of the pure passage of time in terms of the first change. If I tell you that Bikila [the 1960 Olympic marathon champion] is running at the rate of twelve miles per hour of the pure passage of time, for example, then I have also told you that the pure passage of time is flowing at the rate of one hour for every twelve miles run by Bikila. (p. 842)

I must admit to being confused by this passage. I take it that the standard procedure for determining the speed at which Bikila is running is this one:

Pick two times S and T one second apart; find the number r that represents the distance in meters that Bikila has run between S and T; then his average speed between S and T is r meters per second.

Markosian says that at the end of this procedure we get the speed at which Bikila is running with respect to the "pure passage of time." Now by "the pure passage of time" Markosian means objective becoming, the robust passage of time that the block universe theory lacks. But believers in the block universe use this very procedure to determine Bikila's speed (with the same result), and they do not think that the procedure's output is his speed with respect to the pure passage of time. This at least suggests that even if the moving spotlight theory is true the use of this procedure does not yield Bikila's speed with respect to the pure passage of time. And if the result of using this procedure is not his speed with respect to objective becoming, then "reversing" the result (taking its reciprocal) is not the rate of objective becoming with respect to Bikila's motion through space.

Of course, it is okay if the standard way of giving Bikila's speed is not his speed with respect to the pure passage of time. Markosian just needs there to be some way, perhaps a non-standard way, of giving his speed so that it is given with respect to the pure passage of time. What might this be?

One natural approach leads to a dead end. Think about it in the context of MST-Supertime (what follows can easily be translated into the terms MST-Supertense and MST-Time use). In MST-Supetime objective becoming has to do with variation in supertime, so a natural thought is that to get Bikila's speed with respect to objective becoming we should look at how far he moves each super-second. In more detail, we should pick two supertimes P and Q, one supersecond apart, and look at the distance between where (in space) Bikila is at P and where he is at Q.

But this procedure makes no sense. Relative to supertime P there is no one place in space where Bikila is. Relative to a single supertime Bikila is at different places at different (ordinary) times. So this natural thought does not go anywhere.

But there is a more complex procedure that does seem to work. Instead of looking at the distance between where Bikila is at supertime P and where he is at supertime Q, look at the distance between where Bikila is at the time that is present relative to P and the time that is present relative to Q. Then divide that distance by one supersecond. The result will be that Bikila's speed is some number of meters per supersecond. (Since time passes at a rate of one second per supersecond, the number giving his speed in superseconds is the same number that gives his ordinary, "block universe" speed in seconds.)

What about Markosian's proposal that the reciprocal of Bikila's speed with respect to the pure passage of time is one of way giving the rate of time's passage? We are supposed to be able to say things like "time passes at a rate of 1/12 of a (super)hour per mile run by Bikila." When you unpack what this means I think it comes out true. For it amounts to saying that it took 1/12 of a superhour for presentness to get from T_1 to T_2, where T_1 and T_2 are times between which Bikila ran one mile.

Still, even if Markosian has given a believer in the moving spotlight theory a way to answer "How fast does time pass?", she should also want other ways to answer it. For one thing, as Van Cleve points out, if Bikila speeds up (runs faster) then someone who adopts Markosian's answer will have to say that the pure passage of time slows down ("Rates of Passage," p. 149). A believer in objective becoming will want a definition of the rate of time's passage that makes it much harder than this for that rate to vary. For another, many believers in objective becoming will want to say that the pure passage of time goes on, and goes on at some rate, even if nothing else changes. But Markosian's procedure only gives a result if there is some ordinary change to compare objective becoming to.

8

The Challenge from Relativity

8.1 Time in Minkowski Spacetime

The special theory of relativity revolutionized our thinking about space and time. And it is commonly thought that joining this revolution requires rejecting the existence of objective becoming. Is this true? Before I can address this question I must introduce the relevant parts of special relativity.[1]

The most important part of the theory is what it says about the geometry of spacetime. According to the special theory of relativity, the geometry of space-time is not the geometry of Newtonian spacetime (the geometry I described back in Chapter 2). As a way of getting to what the geometry of spacetime is let me review some facts about Newtonian spacetime.

The basic geometrical concepts in Newtonian spacetime are the concept of spatial distance and the concept of temporal distance. The geometry of Newtonian spacetime can be described using a set of axioms involving these concepts, just as the geometry of Euclidean space can be described using a set of axioms involving the concept of spatial distance. It is by using these concepts that we can define in that geometry "point of space" and "instant of time."

Let me remind you of those definitions. Pick any two spacetime points. The spatial distance between them is either zero or some positive number (of, say, meters); the temporal distance between them is either zero or some positive number (of, say, seconds). Then the definitions are

(D1) Spacetime region R is an instant of time $=_{df}$ every point in R is at zero temporal distance from every other point in R, and every point at zero temporal distance from a point in R is (already) in R.

---

[1] I am only going to say enough about the geometric structure of Minkowski spacetime to discuss the conflict between relativity and objective becoming. There are lots of great books that are more comprehensive. Two are *General Relativity from A to B* by Robert Geroch and *It's About Time: Understanding Einstein's Relativity* by N. David Mermin.

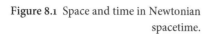

Figure 8.1 Space and time in Newtonian spacetime.

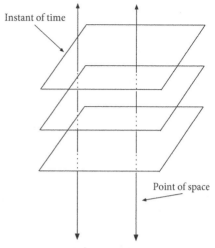

(D2) Spacetime region R is a point of space =$_{df}$ every point in R is at zero spatial distance from every other point in R, and every point at zero spatial distance from a point in R is (already) in R.

These definitions generate a picture of how Newtonian spacetime divides up into points of space and instants of time (Figure 8.1, which reproduces Figure 2.6).

(As an aside, I have also been assuming, and will continue to assume, that there is a distinction between the past and the future directions in time. For any two points in spacetime that are some positive temporal distance apart, we can say which one is later that the other. Where does this distinction come from? Many believers in the block universe deny that facts about which points are later than which others—facts about temporal precedence—are facts about the geometrical structure of spacetime. Instead they are (merely) facts about how matter is arranged in spacetime. An alternative is to say that the basic geometrical concepts also include the concept of temporal precedence. I am going to ignore this debate here.)

According to the special theory of relativity, the geometry of spacetime is the geometry of Minkowski spacetime. The geometry of Minkowski spacetime is very different from that of Newtonian spacetime. In Minkowski spacetime it makes no sense to ask for the spatial or temporal distance between two spacetime points. (More accurately, it makes no sense to ask for them "absolutely," not relative to any "frame of reference." But more on this later.) The concept of the "spacetime interval" between two spacetime points replaces the concepts of spatial and temporal distance as the basic geometrical concept. So what is the

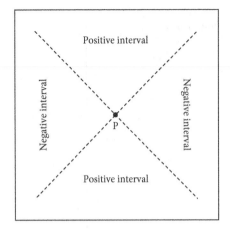

Figure 8.2 The structure of Minkowski spacetime.

spacetime interval, and what does it have to do with the more familiar notions of spatial and temporal distance? That will take some explaining.

Unlike spatial and temporal distances in Newtonian spacetime, the spacetime interval between two points in Minkowski spacetime can be negative, not just zero or positive. If you pick a point P and look at the spacetime intervals between P and other points in Minkowski spacetime you get the picture in Figure 8.2. The points that are interval zero from P form a double-cone. (In a two-dimensional diagram they form an X. Rotate the diagram in Figure 8.2 horizontally around P to get a three-dimensional diagram; the X traces out a double-cone as it rotates.) The interval between P and points inside the cones is positive. The interval between P and points outside the cones is negative.[2]

The most direct connection between the spacetime interval and the more familiar notions of spatial and temporal distance is its connection to temporal distance. Think back to how we represented the spatiotemporal career of, say, a person, in Newtonian spacetime. We represented it with a line in a spacetime diagram (as in Figure 8.3; I use "line" to cover curved and straight lines). This is the person's "worldline." But not just any line is eligible to be a worldline of a person (or some other, relatively small, massive material body). To represent the spatiotemporal career of a massive body the line cannot be tangent to an instant of time.[3] For the reciprocal of the slope of the tangent is supposed to represent

---

[2] It is a matter of convention whether we say that the interval between P and the points inside the cones are positive or negative. What matters is that the intervals for the points inside the cones and the points outside the cones have opposite signs.

[3] For the mathematically sophisticated readers I should say that here and throughout I limit my attention to differentiable lines, lines that have tangents at every point. Also when I speak of regions of spacetime I mean to limit my attention to "nice" regions, regions that are also (differentiable) sub-manifolds.

the body's speed (that is, its ordinary speed through space). Vertical lines, which have "infinite slope," represent bodies at rest. Lines with tangents that have very small slopes represent bodies moving very fast. If the worldline of some body ever became tangent to an instant (in our two-dimensional spacetime diagrams, tangent to a horizontal line) it would represent the body as moving infinitely fast, which is impossible.

Now suppose that the person whose worldline is drawn in Figure 8.3 carries a (normal, functioning) stopwatch, which he starts at spacetime point P. What will the watch read at Q? In the physics of Newtonian spacetime the answer is simple. It will read a number equal to the absolute temporal separation between P and Q.

In Minkowski spacetime bodies still have wordlines but the rules are different. It is still true that worldlines of massive bodies cannot ever have tangents that are horizontal in our two-dimensional diagrams. But in Minkowski spacetime there is another, stronger, restriction on possible worldlines of massive bodies. These worldlines must stay "inside the cones." That is, if the worldline of a massive body passes through a point P then that worldline is coming in to P from inside one lobe of the double-cone at P and heading out of P into the other lobe. (Figure 8.4 illustrates this restriction.) Lines that satisfy this condition are called timelike lines.

To see what this restriction means physically we need another definition. If, for every point P on line L, the straight line that is tangent to L at P lies on the special cones in spacetime, then we say that L is a lightlike line. They have this name because the lightlike lines are the possible worldlines of massless particles, like photons. (The two special cones at a point P are called its light cones.) If

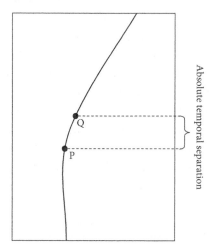

Figure 8.3 A worldline in Newtonian spacetime.

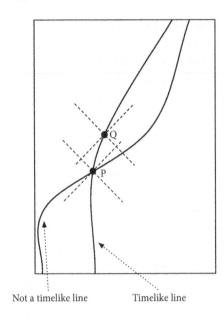

Figure 8.4 Worldlines in Minkowski spacetime.

Not a timelike line   Timelike line

the worldline of a massive body did not stay inside the cones that body would move as fast as, or faster than, light. Thus, the restriction that worldlines of massive bodies must be timelike corresponds to the fact that, in the special theory of relativity, massive bodies must travel slower than the speed of light.[4]

By the way, in speaking of a worldline "coming in" and "going out" of a spacetime point I am assuming that in Minkowski spacetime there is a distinction between the future and the past directions in time. But the way this distinction works in Minkowski spacetime is different from the way it works in Newtonian spacetime. In Newtonian spacetime the assumption that there is a distinction between the past and future directions is the assumption that we can pick any two points and say whether the first is earlier than, simultaneous with, or later than the second. This is not so in Minkowski spacetime (for one thing, because there is no absolute notion of temporal separation and so no absolute notion of simultaneity). In Minkowski spacetime the assumption is that we can pick any two points *on a timelike or lightlike line* and say whether the first is earlier than or later than the second. This is equivalent to the assumption that we can pick any point and say which of the light cones at that point is the future light cone and which is the past light cone. These assumptions are represented in spacetime

---

[4] At least, "ordinary" massive bodies do. Many philosophers believe that special relativity does not in fact prohibit faster-than-light travel, even if no material things in our universe move this fast. See Arntzenius, "Causal Paradoxes in Special Relativity" for discussion.

diagrams the way you expect. Later points on a given timelike or lightlike line are closer to the top of the diagram and future light cones open toward the top of the diagram.

So even though there is no notion of absolute simultaneity in Minkowski spacetime there is a notion of absolute temporal precedence. The points that are absolutely earlier than P are just the points on or inside P's past light cone. This notion of absolute temporal precedence cannot be defined in terms of the spacetime interval. It is an extra bit of geometrical structure.

Let's get back to how temporal separation is defined (when it is defined) in Minkowski spacetime. Suppose again that someone starts his stopwatch at P and looks at it at Q. What in the physics of Minkowski spacetime determines what the stopwatch reads? Not the absolute time separation between P and Q. That is undefined. But there is a time separation between P and Q relative to that person's worldline. That is what the watch will read. The time separation between P and Q relative to someone's worldline is called that person's proper time between P and Q.

Proper time is defined using the spacetime interval, and is defined relative to any timelike line, whether it is the worldline of some massive body or not. The definition is easiest to state for straight timelike lines:[5]

> The temporal separation between P and Q relative to a straight timelike line that passes through both is equal to $\sqrt{I(P, Q)}$, the square-root of the spacetime interval between P and Q (see S's worldline in Figure 8.5).

For curved timelike lines divide the part of the line between P and Q into many small parts with endpoints P_1, P_2, ..., P_n. Then the temporal separation, for this person, between P and Q is approximately equal to the sum

$$\sqrt{I(P, P_1)} + \sqrt{I(P_1, P_2)} + \cdots + \sqrt{I(P_n, Q)},$$

and is exactly equal to the limit of this sum as n goes to infinity. Note that this definition makes sense only if $I(X, Y)$ is never negative, which it can never be for a timelike line.

---

[5] One of the dangers one faces when reading a spacetime diagram is that of assuming that the geometry of the piece of paper on which the diagram is drawn matches the geometry of the space-time the diagram represents. It is certainly true that one line in Figure 8.5 is straight and the other is curved. It does not automatically follow from this that there is a distinction between straight and curved lines in the geometry of the spacetime the diagram represents. In fact there are some space-time geometries in which there is no distinction between curved and straight worldlines. But there is in Minkowski spacetime.

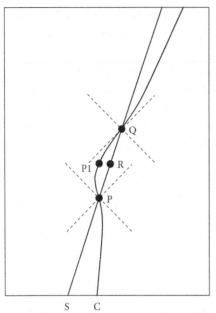

Figure 8.5 Proper time in Minkowski spacetime.

Relative temporal separation: proper time from P to Q is less for C than for S.

It turns out that in Figure 8.5 the spacetime interval between P and P1 is smaller than that between P and R, and the interval between P1 and Q is smaller than that between R and Q. In general it is strange but true that someone whose worldline is curved between P and Q records a shorter elapsed (proper) time between P and Q than someone else whose worldline is straight between P and Q. This is the "twin paradox." Twins who depart from the same location but travel different paths through spacetime may have aged by different amounts when they meet up again. This fact illustrates one way in which using the Euclidean geometry of a piece of paper to represent the geometry of Minkowski spacetime can be misleading. The curved line is longer in the diagram even though it represents a line of shorter temporal length in spacetime.

An important fact about this definition of temporal separation in Minkowski spacetime is that it only defines the temporal separation of two spacetime points relative to a timelike line that passes through both. It only defines the temporal separation between events that happen "right in front of you." What about the temporal separation, relative to a timelike line, between events that do not both fall on that line?

There is a way to extend the definition to cover that case, when the line is straight. Figure 8.6 depicts the relevant situation. A straight timelike line *w* passes through P but not Q. Here is the definition:

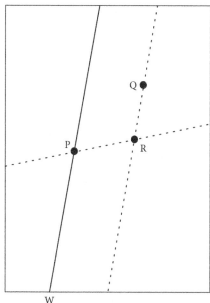

Figure 8.6 Relative temporal separation in Minkowski spacetime.

The temporal separation, relative to *w*, between P and Q is equal to $\sqrt{I(R, Q)}$, where R is the point where the (timelike) line through Q parallel to *w* crosses the line (or, in more than two dimensions, the hyperplane) through P that makes a "relativistic right angle" with *w*.

This definition assumes that for any line in Minkowski spacetime and any point not on it there is a unique line passing through the point parallel to the line. And it uses the notion of a relativistic right angle. But I have not proven the existence and uniqueness of parallel lines or defined "relativistic right angle." I am not going to do these things because I do not think that seeing the proof or the definition is necessary for what follows. So let me just say this. It is safe when discussing parallelism to "read off" facts about parallelism in spacetime from facts about parallelism in a spacetime diagram. That is, two lines are parallel in spacetime if and only if the lines on the diagram that represent them are parallel in the diagram. But it is not safe to do the same for right angles. Lines that are at right angles in the diagram are not necessarily at right angles in spacetime, and vice versa. To check whether intersecting lines meet in a relativistic right angle, look to see whether the angle in the diagram by which one of the lines is inclined from the vertical is equal to, but in the opposite direction from, the angle by which the other line is inclined from the horizontal (as in Figure 8.7).

Let me make two small comments about this definition of temporal-separation-relative-to-a-line before I explain where it comes from. First, the

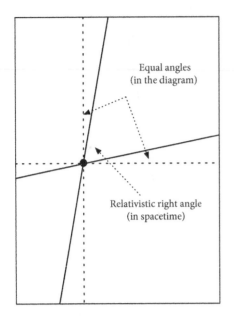

Figure 8.7 Relativistic right angles.

Equal angles
(in the diagram)

Relativistic right angle
(in spacetime)

definition requires that the line pass through one of the points in question. But this is clearly not essential. If it passes through neither P nor Q just consider another straight timelike line, parallel to the initial line, that does pass through P, and apply the definition with that line.

My second remark is about terminology. Let us say that a frame of reference is a thing relative to which any two spacetime points have a spatial and temporal distance. Then a straight timelike line defines a frame of reference. (I have only given the definition of relative temporal separation, but there is a similar one, using the interval, for relative spatial distance. Note that distinct lines will define the same frame of reference if and only if they are parallel.) Relative to a given frame of reference spacetime looks Newtonian. But "absolutely" speaking, "outside" of any frame of reference, the Newtonian notions of spatial and temporal distance make no sense.

The geometrical definition of relative temporal separation I have been discussing may seem mysterious. What does this condition stated in terms of relativistic right angles have to do with our ordinary notion of temporal separation? I have already answered this question when P and Q lie on a single worldline. Temporal separation as so defined is what the watches we carry with us measure. So to have a complete answer it is enough to answer it in the case where P and Q do not lie on a single worldline.

Start with the case where the definition says that the relative temporal separation between P and Q is zero—when P and Q are simultaneous relative

Figure 8.8 Relative simultaneity in Minkowski spacetime.

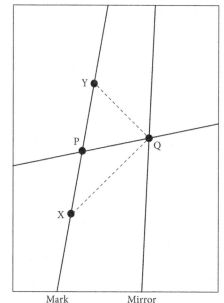

Mark   Mirror

to *w*. Figure 8.8 shows the situation. Someone's—say, Mark's—worldline passes through P but not Q. The line through P at a relativistic right angle to Mark's worldline passes through Q. Why is it a good idea to define relative simultaneity so that P and Q count as simultaneous relative to Mark's worldline?

Suppose that there is a mirror that passes through Q. Suppose that Mark is carrying a flashlight. Suppose he aims it toward the mirror and turns it on at spacetime point X. The photons from the flashlight head toward the mirror, hit the mirror at Q, reflect off the mirror, and head back to Mark. The first returning photons hit his eyes at Y. Suppose he timed the whole process and found that the proper time (for him) between X and Y was some number of seconds N. (Here we are talking about the temporal separation between two points that lie on a single timelike line, which we already understand.) Then it is quite natural for Mark to assume that the photons hit the mirror half-way through their round-trip,[6] and so that the spacetime point on his worldline that is

---

[6] The mirror is moving relative to Mark. Now when you throw (say) a tennis ball the speed at which it bounces off a racquet depends on the speed with which that racquet is moving. So if you throw a tennis ball at a moving racquet it is wrong to assume that the ball moved at the same speed on its way out as on its way back. Hence it is wrong to assume that the ball hit the racquet half-way through its round-trip. Does this show Mark is wrong to assume that the photons hit the mirror half-way through their trip? No, because photons are not like tennis balls. Whatever the state of motion of the mirror (or the flashlight), photons travel on the light cone.

simultaneous with Q is the point $N/2$ seconds later than X and $N/2$ seconds earlier than Y.[7] And that is point P.[8]

The general case (depicted back in Figure 8.6), where Q is any point not on w, is easy to make sense of using the two special cases we already understand. The definition says that the temporal separation, relative to w, between P and Q is equal to the temporal separation, relative to w, between Q and R, where (i) P and R are simultaneous relative to w, and (ii) Q and R both lie on a straight timelike line that defines the same frame of reference as w.

8.2 Preview

I have said only a little bit about the geometry of Minkowski spacetime. There is a lot more to say. (For example, I have not said anything about the connection between the spacetime interval and spatial distances relative to a frame of reference). But I have said enough to start thinking about the nature and extent of the conflict between the moving spotlight theory and the special theory of relativity.

The best that an opponent of objective becoming could hope for is that the moving spotlight theory is inconsistent with special relativity. But right away there is a problem. There is some controversy over what exactly the special theory of relativity says. For this reason I think it is best to focus on whether a moving spotlight is consistent with the most uncontroversial part of the theory, namely the thesis that

---

[7] I said that it is quite natural for Mark to assume this. There has been a debate about whether there are other assumptions Mark might make that lead to "non-standard" definitions of relative simultaneity. Why not assume that the photons hit the mirror one-third of the way through their round-trip? I think the upshot of those debates is this. Of course Mark can mean whatever he wants to by "relative simultaneity," including things other than the meaning I have been discussing. But if you want a definition of "relative simultaneity" in Minkowski spacetime such that (i) the only geometrical notion used in the definition is the spacetime interval, (ii) "X and Y are simultaneous relative to frame F" is an equivalence relation, and (iii) each equivalence class intersects each timelike line exactly once, then this is the only possible definition. A more precise statement, and proof, of this claim may be found in Giulini, "Uniqueness of Simultaneity"; see also Malament, "Causal Theories of Time and the Conventionality of Simultaneity."

It is important to separate the question of how relative simultaneity is to be defined in Minkowski spacetime and the question of which spacetime geometry to use for our physical theory in the first place. Sometimes the claim that the definition of relative simultaneity is conventional just amounts to the claim that a physics in Minkowski spacetime is not the only one consistent with the experimental evidence.

[8] I have not proved that that is point P. Certainly the distance in the diagram between X and P looks to be the same as that between P and Y, but we have seen that diagram-distance does not always correspond to "distance" (either spacetime interval or relative temporal separation) in spacetime. In this case, however, it does. (That is because two pairs of points that are the same distance apart in the diagram and also lie on parallel lines have the same spacetime interval.)

(1) The geometry of spacetime is the geometry of Minkowski spacetime.

So let us consider the claim that the moving spotlight theory is inconsistent with (1). How bad would it really be for the believer in objective becoming if this were true? Well, the argument might go, we have all sorts of excellent experimental evidence for (1). And, the argument continues, it is irrational to be aware of that evidence, and of its support for (1), while believing anything inconsistent with (1).

How good is this argument? One problem is that in fact we have lots of experimental evidence that (1) is false. (For example, the evidence that supports the general theory of relativity.) But set that problem aside for now. I want to start with the argument's central claim. Is it really true that the moving spotlight theory is inconsistent with (1)?

It is true, or at least close to being true. But its being true is not worth much, even if we grant that it is irrational to believe anything inconsistent with (1). What would be really bad is if it were impossible to devise new versions of the moving spotlight theory that are not inconsistent with (1). Of course a version of the moving spotlight theory that is consistent with (1) is interesting only if it remains true to the spirit of the classical moving spotlight theory and the ideas that motivate it. But I think there are relativistic versions of the moving spotlight theory that meet these conditions.

If the moving spotlight theory is inconsistent with (1) then perhaps the theory can be refuted. (Though, again, perhaps not; I will return to this below.) But it is not inconsistent with (1). Relativity does make describing objective becoming more complicated. Maybe in the end this "counts against" the existence of objective becoming. When objective becoming is weighed against the block universe theory the complications in the moving spotlight theory required to accommodate relativity might be a chip in the block universe theory's pan. But I think it is a small chip that makes no difference to the outcome. Those who reject objective becoming tend to be "hard-headed scientific types" who would have rejected it even if they had never heard of the theory of relativity. They may of course regard the difficulties that come with relativity as extra reasons for rejecting it; and finding more reasons to reject it can be an interesting intellectual pursuit. But, I think, the typical believer in objective becoming is not going to give up their belief in the face of the relativistic difficulties.

8.3 Inconsistency

I am getting ahead of myself. We need to see what these new versions of the moving spotlight theory look like. But before we do that we should see how the versions I have discussed are inconsistent, or something like inconsistent, with (1). In the last few chapters I have focused on two versions of the theory, MST-Supertense and MST-Time. But it is easiest to focus this investigation on MST-Supertime. (I will return eventually to the others.) Here again is one of the basic claims of MST-Supertime:

(2) Relative to each point in supertime exactly one instant of time is present. Each instant is present relative to some point in supertime, and later instants are present relative to later points in supertime.

The theory also says that presentness moves continuously at a constant rate.[9] But we can ignore these parts of the theory here (and throughout this section). Now whenever I have drawn pictures illustrating MST-Supertime I have presupposed the 3 + 1 view of spatiotemporal reality. But when discussing relativity we really need to take the 4D spacetime point of view seriously. So we need a picture that shows the spotlight in spacetime. Figure 8.9 illustrates MST-Supertime from this point of view. An instant of time is just a region of spacetime, and in the figure I have indicated which region is present relative to a point in supertime with a solid line.

Supertime

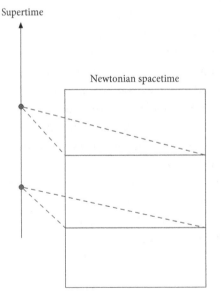

Newtonian spacetime

Figure 8.9 The spotlight in Newtonian spacetime.

[9] Of course the official statement of the theory in fundamental vocabulary does not use the word "presentness."

Whether MST-Supertime is inconsistent with (1) depends on what "instant of time" means as it occurs in (2). So let's assume that it has the same meaning it had when we were discussing Newtonian spacetime, the meaning given in definition (D1). Here is that definition again:

(D1) Spacetime region R is an instant of time $=_{df}$ every point in R is at zero temporal distance from every other point in R, and every point at zero temporal distance from a point in R is (already) in R.

Now the key fact about Minkowski spacetime is that in it there is no such notion as absolute temporal distance. But this notion appears in (D1). On some views this means that "instant of time"—and therefore (2)—is meaningless if (1) is true. That is not the same as being inconsistent with (1), but it is close enough.

I am not sure whether it is right that (2) is meaningless if (1) is true. Whether it is depends on how certain questions in the philosophy of language are to be answered.[10] But if (2) is not meaningless then it is straightforwardly inconsistent with (1). If it makes sense to ask which regions of Minkowski spacetime are instants the answer will be "none of them." But (2), if meaningful at all, entails that there are instants of time.

---

[10] It may be that any predicate that is by definition equivalent to an open sentence containing a meaningless expression (here, "temporal distance") is itself meaningless. Alternatively, it may be that such a predicate is meaningful but that nothing can satisfy it.

9

Relativity and the Passage of Time

9.1 Neo-classical Spotlight Theories, I

At the end of the last chapter I went through an argument that the moving spotlight theory is inconsistent with (1),[1] a core commitment of the special theory of relativity. Some believers in objective becoming shrug their shoulders at this conclusion. They respond with this line of thought:[2]

The reason that there are no instants of time in Minkowski spacetime is that there is no way to define "P and Q are (absolutely) simultaneous" (or "the temporal separation between P and Q is zero") using just the spacetime interval. (No way to define it, that is, so that it behaves formally as it should. No way to define it so that it is reflexive, symmetric, transitive, and non-trivial.) So when my opponent says that the moving spotlight theory is inconsistent with special relativity what he really means is that the moving spotlight is inconsistent with the claim that all there is to the geometry of spacetime is the spacetime interval. But I am happy to reject that claim. Part of the geometry of spacetime is captured by facts about the spacetime interval; but not all of it. To capture the rest we need one of the basic geometrical notions of Newtonian spacetime, namely the notion of (absolute) temporal separation.

Does this mean that I reject the special theory of relativity? I certainly do not think that all those experimental tests of the theory were flawed. I do not think the physicists' equipment was broken or that they mis-read where some pointer was pointing. What the experiments that favor the special theory of relativity show (I am thinking of experiments like the Michelson–Morley experiment) is that no experiment can determine whether the temporal separation between two distant events is zero or not. But the history of philosophy shows that it is bad epistemology to refuse to believe something just because there is no way to set up an experiment that will help us figure out whether it is true.

This much is certainly true: if the physicists tell us their experiments support the hypothesis that spacetime has a certain geometrical structure, we should not believe that spacetime has less structure than that. But it is going much farther to say that we also should not believe that spacetime has more structure than that. This stronger

---

[1] The numbering of sentences in this chapter continues the numbering in the previous one.
[2] Philosophers who endorse something like this line of thought include Prior, in "Some Free Thinking about Time," and Markosian, in "A Defense of Presentism," though they are defending presentism, not the moving spotlight theory.

claim is false. There might be reasons to believe that spacetime has geometrical structure beyond the structure that the physicists say they need, reasons that physicists are neither interested in nor specially qualified to investigate.

Indeed, if the objection is that it is unreasonable to believe that spacetime has more properties than the geometrical properties that physics requires then the special theory of relativity is the least of our worries. Ignore over a hundred years of physics and pretend that it is Newtonian spacetime that the physicists like. Even then the moving spotlight theory "adds something." It adds facts about which time is present. Taking Minkowski spacetime and adding facts about which points have zero temporal separation hardly seems worse than this.[3]

I do not have much to say against the general philosophical viewpoint expressed in the second to last paragraph of this speech. (That does not mean I accept it.) But I do not think it applies to the case at hand. For there are versions of the moving spotlight theory that do without a notion of absolute simultaneity. And the best arguments for the moving spotlight theory, the arguments that physicists are "not specially qualified to investigate," do not favor the versions that add a notion of simultaneity over the versions that do not. In fact I think the opposite is true. Those arguments favor the versions without absolute simultaneity.

Before I can defend these claims we need to take a closer look at versions of the moving spotlight theory that add a notion of absolute simultaneity to Minkowski spacetime. The discussion so far has presupposed that these versions of the theory are inconsistent with (1). In fact, however, not all of them are.

Let me start with versions of the moving spotlight theory that are certainly inconsistent with (1). These theories begin by "directly" adding a notion of absolute simultaneity to the geometrical notions used to characterize Minkowski space-time. Call theories that do this type-G theories ("G" for "geometrical"). A type-G theory assumes that "the temporal separation between P and Q is r" is a meaningful piece of fundamental geometrical vocabulary along side "the spacetime interval between P and Q is r."[4] And it says that the set of regions of spacetime R such that R contains all and only the points that are at zero temporal separation from some given point—the set of simultaneity-slices of spacetime—partitions spacetime in the way familiar from Newtonian spacetime. (The theory may also

---

[3] Zimmerman in "The Privileged Present" endorses this last idea, again in the context of presentism.

[4] Actually, a believer in the block universe may not want to say that any of these predicates belong to the fundamental vocabulary for describing the geometry of spacetime. (One reason is that these predicates correctly describe the geometry of spacetime only if there are numbers, and one might want a way of describing that geometry that could be true even if numbers did not exist.) No matter. What is important here is the claim that "temporal separation" is as fundamental as "spacetime interval." Shortly I will describe a theory in which it is not.

say that the set of simultaneity-slices coincides with the set of simultaneity-slices-relative-to-F, for some inertial frame of reference F.[5]) Then claim (2) from Chapter 8 (repeated here with the same number) can continue to be part of a statement of MST-Supertime, with the meaning of "instant" given by (D1):

(2) Relative to each point in supertime exactly one instant of time is present. Each instant is present relative to some point in supertime, and later instants are present relative to later points in supertime.

(D1) Spacetime region R is an instant of time $=_{df}$ every point in R is at zero temporal distance from every other point in R, and every point at zero temporal distance from a point in R is (already) in R.

That is what a type-G theory looks like. Type-G theories are inconsistent with (1). But there is another way to add a notion of simultaneity to Minkowski spacetime. Pick two points of spacetime P and Q such that the straight line through P and Q is timelike. Let F be the frame of reference determined by this line. Then define the term "F-instant" like this:

Spacetime region R is an F-instant $=_{df}$ every point in R is at zero temporal distance relative to frame of reference F from every other point in R, and every point at zero temporal distance relative to frame of reference F from a point in R is (already) in R.

Let us say that F-instant A is absolutely earlier (/later) than F-instant B if and only if every point in A is in the past (/future) light-cone of some point in B. Then each F-instant will be either absolutely earlier than or later than any other F-instant. Using the notion of an F-instant we can formulate a version of MST-Supertime that says (see Figure 9.1):

(3) Relative to each point in supertime exactly one F-instant is present. Each F-instant is present relative to some point in supertime, and later F-instants are present relative to later points in supertime.

This theory differs from type-G theories over the status of the predicate "absolutely simultaneous." Statements of type-G theories, like (2), presuppose the meaningfulness of this predicate. Statement (3), by contrast, does not. "F-instant"

[5] "Simultaneity-slice-relative-to-F" means the same as "F-instant," which I define in the next paragraph. Another way to put the claim I make here in the text is to say that the simultaneity-slices form a foliation of spacetime by instants. I will explain this bit of terminology in Section 9.2. I say the theory "may" say that the set of simultaneity slices is the set of F-instants for some F because it may, instead, say that that set is a set of instants on a more liberal sense of "instant," which I give as definition (D3) below.

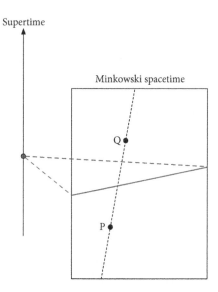

Figure 9.1 One spotlight in Minkowski spacetime.

is not defined using "absolutely simultaneous." It is defined using the spacetime interval, the notion of temporal precedence, and names of particular spacetime points (P and Q). So theories that contain (3), or something like (3) but with a name for a different frame of reference inserted, instead of (2), can use "present" to define "absolutely simultaneous":

- X and Y are absolutely simultaneous $=_{df}$ there is some point in supertime relative to which X and Y are both present.

Call theories that add simultaneity this way, by defining it in terms of presentness, type-N theories ("N" for "non-geometrical"). The notion of absolute simultaneity in a type-N is not a geometrical one. So type-N theories do not add any geometrical structure to Minkowski spacetime. They are consistent with (1).

Is there any reason to prefer type-N theories to type-G theories? Maybe so. Go back to the speech I put in the mouth of the defender of objective becoming. The speech ended with this idea (this is a re-statement, not a quotation):

Even classical theories of objective becoming were in the business of "adding" something to the block universe. A theory that adds something to Minkowski spacetime is not doing anything different and so is not doing anything worse.

But only type-N theories do "the same thing" that the classical spotlight theory does. Only these spotlight theories say that the geometry of spacetime is just as it is in the (classical or relativistic) block universe.[6] Only these spotlight theories disagree with the block universe only over whether there is any such extra- or non-geometrical feature of spacetime as presentness. Type-G theories, by contrast, say that there is more to the geometry of spacetime than the block universe says. The classical moving spotlight theorist does not say anything like this.[7]

How much does all this really matter? So what if type-G theories depart from the block universe in a new kind of way? Does that really make them worse off, when we take relativity theory seriously, than the classical version of the moving spotlight theory is, when we assume classical physics?

Maybe. But type-N theories face difficulties too. Understood in a certain way, type-N theories are inconsistent with a certain principle, which may be roughly stated as

(4) The laws do not treat differently regions of spacetime that are geometrically the same.[8]

If a type-N theory says that something of the form (3) is not just true but also a metaphysical necessity or a law of metaphysics then it is not consistent with (4). For no frame of reference is geometrically unique. But if something of the form (3) is true then one frame plays a special role in the "laws of metaphysics."

Many philosophers have argued that physical theories that are inconsistent with (4) are defective in some way. One might make a similar argument that metaphysical theories, like type-N theories, that are inconsistent with (4) are also defective in some way. This argument, however, has weaknesses. Must metaphysical theories be bound by rules that apply to physical theories? Even if they must, not all type-N theories are inconsistent with (4). As I said above, a theory that contains (3) is inconsistent with (4) only if it says that (3) is something like a law of metaphysics. A type-N theory may say that (3), while true, is contingent. In

---

[6] Well, they may disagree about whether the distinction between the past and future directions in time is a geometrical one. But the transition from classical to relativistic physics makes no difference to this disagreement.

[7] Another approach to reconciling objective becoming with relativistic physics does not add a notion of simultaneity to the geometry of Minkowski spacetime, but instead says that the geometry of spacetime is the geometry of Newtonian spacetime. (Craig takes this approach in *The Tenseless Theory of Time*.) This approach is relevantly similar to type-G theories. Both tinker with the geometry that fits best with the relevant physics. (I do not yet include quantum mechanics in the relevant physics; I will say something about it below.)

[8] This statement of the principle is not very precise. A more precise statement is: every symmetry of the spacetime geometry is a symmetry of the laws. In Minkowski spacetime this means that the laws are Lorentz invariant. John Earman made heavy use of a principle like this in *World Enough and Space-Time* (section 3.4). The status of principles like this has received a lot of discussion since; see Belot, "Symmetry and Equivalence," for a recent and somewhat skeptical evaluation of them.

a theory like that, the foliation that contains the regions that are present varies from possible world to possible world.[9]

Where are we? We may have here some reasons why type-G theories and type-N theories are bad. While I do not think the reasons are very strong, I think the moving spotlight theory can do better. There is a way to formulate the theory so that it is consistent with (1) and lacks the weaknesses of type-N theories. But before I explain this version of the theory I have a few further remarks I want to make about type-N theories.

9.2 Further Notes on Type-N Theories

In 1967 Hilary Putnam argued in his paper "Time and Physical Geometry" that special relativity refutes the "man on the street's" view of time. Now it is unclear just what theory of time Putnam takes himself to be arguing against. He states it as the view that "all and only things that exist now are real." That sounds like presentism. But in other places it looks like his target is the moving spotlight theory. Anyway, his argument did not use just (1) as a premise. The other key premise was a principle he called "There are no Privileged Observers." Exactly what this principle is supposed to be is not entirely clear in Putnam's article,[10] but the idea behind it is close to (4). Since in Minkowski spacetime all frames of reference are geometrically indiscernible, one's theory of time should not—as type-N theories do—favor one frame of reference over another. But, again, this idea is one that believers in objective becoming will have no qualms rejecting.[11]

The type-N theories we have seen pick out the foliation along which presentness moves directly by name. If the theory picks it out by description, rather than by name, then we get a theory that is consistent both with (1) and with (4)

---

[9] Zimmerman, in "Presentism and the Space-Time Manifold," entertains both responses; Forrest favors the second in "Relativity, The Passage of Time and the Cosmic Clock."

[10] Howard Stein's paper "On Einstein–Minkowski Space-time" is a well-known and excellent discussion of Putnam's argument and the interpretive challenges it presents. I will have more to say about Stein's paper.

[11] Lots of philosophers who claim to show that objective becoming is inconsistent with special relativity do what Putnam does. They use in their argument a premise that is neither uncontroversially a part of special relativity nor a claim that their opponents need feel bound to accept. Many follow Putnam and use a premise like principle (4). (Zimmerman lists several in footnotes 78 and 79 of "Presentism and the Space-Time Manifold.") For another kind of example, Saunders's main conclusion, in his paper "How Relativity Contradicts Presentism," is that if presentism is true then "even if one knew all that there is to know, consistent with special relativity, one would not be able to say what [region is present]" ("How Relativity Contradicts Presentism," p. 280). But it is not true in general (it had better not be!) that T is inconsistent with X if the conjunction of T and X implies that there are some things that cannot be known. (It is of course true that sometimes when T and X imply that there are some things that cannot be known then the reasonable thing to do is to reject one of T or X. But whether this is right will depend on the details of the case.)

(and the principle that there are no privileged observers). The center of mass of the universe moves along a straight timelike line. A theory might say that the regions that are present are the regions that are instants relative to the frame of reference determined by the timelike line along which the center of mass of the universe moves.

Theories like this are not plausible. It should be possible for the spotlight to move in an otherwise empty universe. And the path the spotlight takes through spacetime should not depend on the way matter is distributed in spacetime.[12]

Finally, I want to state the most general form of a type-N theory. So far we have seen just one type-N theory. It says that the regions that are present are the instants relative to a particular frame F. Clearly for any frame of reference there is a type-N theory that says that the regions that are present are the instants in that frame. But type-N theories are not limited to theories that say that the regions of spacetime that are present are instants-relative-to-a-given-frame-of-reference. It is true that these kinds of regions resemble the regions that are instants in Newtonian spacetime in certain respects. But there are other kinds of regions of Minkowski spacetime that resemble Newtonian instants in other respects, and there are versions of the moving spotlight theory that say that those kinds of regions are present. It is worth taking a look at them, and seeing which of those theories we should take seriously and which we should dismiss out of hand.

The notion of a timelike line has been important so far:

- A line in spacetime is timelike if and only if it stays inside the light cones at every point it passes through.

There is also a name for lines that satisfy the "opposite" condition. They are the "spacelike" lines:

- A line in spacetime in spacelike if and only if it stays outside the light cones at every point it passes through.

The line in Figure 9.2 is spacelike.

Now if you think back to Newtonian spacetime there are analogues of these notions. If we were to define "timelike line" in Newtonian spacetime so that all possible worldlines of massive bodies were timelike, the definition would say that timelike lines are lines that are never tangent to an instant of time. So in

---

[12] Saunders, in "How Relativity Contradicts Presentism," rejects theories like this for similar reasons. Zimmerman describes several more of them and defends them in "Presentism and the Space-Time Manifold." Maudlin briefly discusses when there is and is not a unique frame of reference picked out by the center of mass of the universe in *Quantum Non-Locality and Relativity* (pp. 186–7).

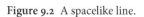

Figure 9.2 A spacelike line.

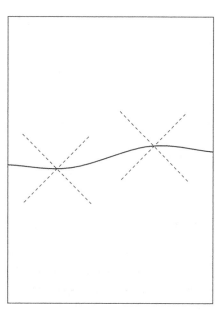

Newtonian spacetime a spacelike line would be a line that is always tangent to an instant. Now a line has this feature if and only if it lies entirely inside a single instant. If we approach this fact from the other direction we get a necessary condition on instants:

> Spacetime region R is an instant of time (in Newtonian spacetime) only if every line that lies entirely within R is spacelike.

This condition is not sufficient. A region satisfying it may fail to be "big enough" (or "wide enough") to be an instant. (A single spacelike line satisfies it without being an instant.) But a second condition that makes it sufficient is not hard to find. Every infinite, straight timelike line should pass through every instant. So we have this equivalence:

> Spacetime region R is an instant of time (in Newtonian spacetime) if and only if every line that lies entirely within R is spacelike, and every infinite straight timelike line passes through R.

Here we have a way of thinking about instants of time in Newtonian spacetime that does not explicitly use the notion of temporal distance. It does use it implicitly, in the Newtonian definition of "spacelike." But "spacelike" has a different definition in Minkowski spacetime, so we can use this way of thinking about instants to generate a definition of "instant" in Minkowski spacetime, a definition that does not falsely presuppose that there is a notion of absolute temporal distance:

(D3) Spacetime region R is an instant of time =$_{df}$ every line that lies entirely within R is a spacelike line, and every timelike line passes through R.

On this definition of "instant" there are instants of time in Minkowski spacetime. In a two-dimensional spacetime the line in Figure 9.2 is an instant of time according to (D3).

Now we can see another way to reconcile the moving spotlight theory with relativity. Let the theory say, as the classical theory does,

(5) Relative to each point in supertime exactly one instant of time is present,

but use (D3) instead of (D1) for the meaning of "instant."[13]

This, however, is not entirely satisfactory. It is consistent with the view as stated so far that a single spacetime point be present from more than one point in supertime. Figure 9.3 illustrates how. From supertime point R one time U that contains P is present; from later supertime point S a different time V is present. But this other time also contains P.

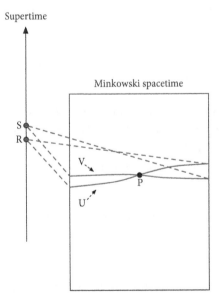

Supertime

Figure 9.3 Present times from two points in supertime.

Minkowski spacetime

---

[13] The availability of this definition of "instant" shows that it is at least misleading to say that the problem the theory of relativity raises for the moving spotlight theory is that there are no instants of time in Minkowski spacetime. In fact, if there is a problem, it is that there are too many. (I will return to this idea.)

It is worth pausing at this point to ask just what is so bad about this. So what if a single spacetime point gets to be present twice? Why isn't it enough that no instant gets to be present twice? The theory may be strange but it is not inconsistent.

The answer is that we want more out of the moving spotlight theory than consistency. All sorts of bizarre metaphysical theories that are not worth taking seriously are consistent. What we want is a version of the theory that is consistent with the ideas that motivate the moving spotlight theory in the first place. I think the best arguments for the moving spotlight theory are arguments that appeal to our experience of the passage of time. (I will discuss them in detail in Chapters 11 and 12.) The best version of the moving spotlight theory will make some connection between which spacetime points (on my worldline) are present and which experiences I am having. But I never return to the same experience twice. Nor does this seem possible unless something truly strange is going on. (Maybe if spacetime were "cylindrical in the temporal direction" it would be possible. But Minkowski spacetime is not.) So barring truly strange circumstances the moving spotlight theory should not permit a single spacetime point to be present at more than one supertime.

Now in the classical version of MST-Supertime this kind of thing is ruled out because (2) contains not just (5) but also

(6) Later instants are present relative to later points in supertime.

We can impose this requirement in Minkowski spacetime too if we give "instant A is later than instant B" a relativistic meaning. I actually already did this for the case where A and B are instants relative to some inertial frame of reference. The definition I gave is this:

F-instant A is absolutely earlier (/later) than F-instant B $=_{df}$ every point in A is in the past (/future) light-cone of some point in B.

Now let us make this definition more general, to apply not just to F-instants but to all instants in the sense of (D3). Then our theory can assert, in place of (5),

(6) Relative to each point in supertime exactly one instant is present. And if supertime Q is later than supertime P, then the instant that is present at Q is later than the instant that is present at Q.

However, (6) still does not quite say what we want the theory to say. Statements (2) and (3) include the claim

(6*) Each instant is present at some point in supertime.

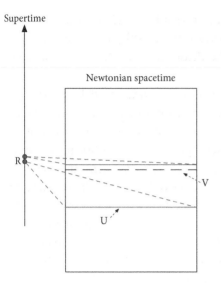

Supertime

Newtonian spacetime

R

V

U

Figure 9.4 Points that are never present.

Statement (6*) is there to ensure that presentness moves continuously into the future. Without (6*) there is nothing to prevent there from being points in space-time that are never present. Figure 9.4 illustrates this kind of situation. Relative to supertime R instant U is present; there is a substantial gap in spacetime between instant U and instant V; but for every point in supertime X later than R, the instant that is present relative to X is later than V. The spacetime points in the gap are never present.

To solve this problem we cannot simply add (6*) to (6). That would make the theory is inconsistent. As Figure 9.3 makes clear, some instants overlap each other.

To solve this problem we can use the concept of a foliation of spacetime by instants. A foliation by instants is a collection of instants (in the sense of (D3)) with the property that each point in spacetime lies on exactly one of them. Foliations by instants certainly exist. For any frame of reference the collection of regions that are instants-relative-to-that-frame is a foliation of spacetime by instants. This is a frame foliation. But not every foliation of Minkowski spacetime is a frame foliation. And foliations of spacetime by instants have just the properties we need. In any foliation each spacetime point is on some instant or other, and each instant is either absolutely earlier than or absolutely later than each other instant.

Now for each foliation of spacetime by instants O we could consider the version of the moving spotlight theory that says

(7) Relative to each point in supertime one instant in O is present. Each instant in O is present relative to some point in supertime, and later instants in O are present relative to later points in supertime.

Now my goal is to write down the most general form of an acceptable type-N theory. By moving from (3) to (7) we admit not just frame foliations but also arbitrary foliations. But with (7) we have over-generalized. Not all instances of (7) are acceptable versions of the moving spotlight theory. Some instances run into trouble when they try to say how fast time passes.

The theory is supposed to say

If P and Q are points in supertime and P is one supersecond later than Q, then the instant that is present relative to P is one second later than the instant that is present relative to Q.

But this only makes sense if there is a way to define the temporal distance between two instants. Now if the instants are instants-relative-to-frame-F then there is no problem. For relative to F any two points have a temporal separation, and moreover any two points from two fixed F-instants have the same temporal separation. So the temporal distance between the instants can be defined to be equal to this temporal separation (relative to F) between points on them.

But this definition of temporal distance between instants does not work in a non-frame foliation. There are several reasons why. Look at the definition again:

(D4) The temporal distance between F-instant I and F-instant J is R $=_{df}$ the temporal distance, relative to frame of reference F, between X and Y is R, where X is any point on I and Y is any point on J.

If I and J are instants in a non-frame foliation then which frame of reference should we use on the right-hand side? And even if we set this problem aside, there is no guarantee that the temporal separation between X and Y is the same, no matter which X we choose from I and which Y we choose from J.

What we need is an alternative to (D4) that works for non-frame foliations. Now I doubt that there is an alternative that works for all non-frame foliations. The thing to do is to restrict our attention to non-frame foliations in which there is a natural way to define the temporal distance between two instants using the geometry of Minkowski spacetime. (I am not sure how to characterize exactly which non-frame foliations these are, or what exactly the alternative to (D4) should be. I will make one suggestion in a footnote.[14]) Once we restrict (7)

---

[14] For any two instants I and J in a given foliation look at all the (not necessarily straight) timelike line segments with endpoints on I and J. Look at the proper times along all of those segments. If

in this way the theories we get will say the right thing about the rate of time's passage. These theories, I claim, are all and only the type-N theories worth taking seriously.

Besides theories that use non-frame foliations by instants in which there is no natural way to define the temporal distance between two instants, what other potential type-N theories am I ruling out? One kind of theory I am ruling out is one that says that the regions that are present make up a foliation of Minkowski spacetime by "back light cones." (Each leaf of the foliation consists of a spacetime point plus the surface of that point's back light cone. Note that this foliation is not a foliation by instants.[15]) We should not take a theory like this seriously. Nor should we take seriously a theory that says that the regions that are present are "bowtie-shaped" (they consist of a single spacetime point plus all points at a negative interval from it). The kind of argument I used against a theory that permits a single spacetime point to be present "more than once" works again here. In fact this kind of argument is going to be increasingly important as this chapter goes on.

Here is how that argument applies to a theory that says that bowtie-shaped regions are present. If a bowtie-shaped region is ever present then an entire segment of someone's worldline (a segment has non-zero temporal length) could be present "all at once," at one point in supertime. But then the theory says that relative to a single point in supertime a continuous series of someone's experiences are all present. This should not be possible.

A similar argument works against the theory that says that the regions that are present are back light cones. In a theory like that it is possible that a photon's entire worldline be present at a single point in supertime. True, a photon cannot have experiences, so the theory does not need to say that a continuous series of someone's experiences are present at a single point in supertime. But I still think this is a decisive objective.[16]

---

none of those proper times is infinite we can try defining the temporal distance between I and J to be their least upper bound.

This definition is adequate for a given foliation if it makes temporal distance additive on that foliation. If I is earlier than J which is earlier than K, then the temporal distance between I and K must equal the sum of the distance between I and J and the distance between J and K. However, the fact that the temporal distances given by the definition exist does not guarantee that temporal distance is additive. It can fail to be additive if the longest timelike line segment from I to K is not made up of the longest segment from I to J and the longest segment from J to K.

If this definition is adequate for a given foliation then that foliation is a candidate to be the foliation along which presentness moves.

[15] It is still true that every point in spacetime lies on exactly one "element" in the foliation.

[16] Hinchliff proposes a theory like this in "A Defense of Presentism in a Relativistic Setting," though his is a relativistic version of presentism, not a relativistic version of the moving spotlight theory.

9.3 Neo-classical Spotlight Theories, II

I have said that believers in objective becoming can do better than type-N theories. The time has come for me to explain why I think this.

Type-N theories face a well-known epistemic problem. If some type-N theory is true, then it is impossible to know that it, rather than some alternative type-N theory, is the true one. But the alternative relativistic theory I will describe does not face this kind of problem.

The best arguments for objective becoming are arguments from our experience of time. It is not yet time to discuss these arguments in detail. For now it is enough to know that these arguments motivate a version of the moving spotlight theory in which there is some connection between which spacetime points are present and which experiences I am having. (I mentioned this before, in Section 9.2.)

But the arguments from experience do not favor one type-N theory over another. Figure 9.5 illustrates this claim. In it we see two different type-N theories. One says that the red (solid) regions are present, the other says that the blue (dashed) regions are present. That is where they differ. But each theory says that exactly one of my experiences is present at each point in supertime. And each theory says later experiences are present at later points in supertime. As far as my experience goes the theories are equivalent. And I have no "access" to other people's experiences. I have no idea which of, say, the President's experiences is present relative to the point of supertime at which this experience (a relatively boring experience as of typing at my computer) is present. So there are no grounds for preferring one of the theories to another.

Figure 9.5 Two neo-classical theories.

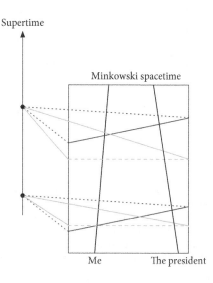

Now you might object that something like this is true even in the classical moving spotlight theory. The classical theory, remember, is set in Newtonian spacetime. There are deviant versions of the classical theory that say that the regions that are present are not instants of time in Newtonian spacetime, but are regions that are, in some sense, "close to" being instants, regions that together make up a foliation of Newtonian spacetime.[17] A theory like that might say, for example, that region R1 in Figure 9.6 is present at some point in supertime, while R2 is present at some later point. Is not this deviant version of the classical moving spotlight theory doing as well as the standard version as far as my experience goes?

Maybe it is doing as well as far as my experience goes. But the theory can still be ruled out. If spacetime is Newtonian then no matter how close R1 is to being an instant it is possible that I be moving so fast that an entire segment of my worldline lies entirely inside region R1. In that scenario the deviant theory says that relative to a single point in supertime a continuous series of my experiences are all present. We do not want a theory that allows this.

In Newtonian spacetime we can use this kind of argument to select a unique foliation of spacetime to be the one along which presentness moves. We cannot

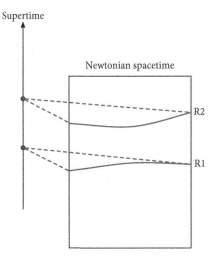

Figure 9.6 A deviant classical theory.

---

[17] It is at the very least not obvious what "close to" could mean here. The first thing to require is that these regions be three-dimensional submanifolds of Newtonian spacetime that intersect every infinite straight timelike line exactly once. Now if these regions are not instants then not all points in them are simultaneous. If they are to be "close" to being instants one might want to require that the maximum temporal distance between any two points in a single region be small. But what does small mean? One second is a relatively short amount of time for us, but much longer for certain insects. There is no absolute sense in which a temporal distance can be small or large.

do the same in Minkowski spacetime. What prevents us is the fact that in relativity theory there is a finite upper limit on speeds. Consider any foliation of spacetime by instants. It is not possible for me to move so fast that an entire segment of my worldline lies entirely inside one of those instants. To do so I would have to move faster than light.

So type-N theories face an underdetermination problem. The source of the problem is that these theories are only partially relativistic. They have made the switch from Newtonian spacetime to Minkowski spacetime. But supertime remains non-relativistic. Call theories like this "neo-classical." In neo-classical theories there is a mismatch between the structure of supertime and the structure of the temporal aspects of Minkowski spacetime.

One way to eliminate the mismatch between the structure of supertime and Minkowski spacetime is to give up the idea that it is entire instants of time that are present. "After all," the thought might go,

if it is fidelity to my experience that provides whatever grounds there are to believe the moving spotlight theory, then all that matters is that my experiences are successively present. A theory can say that if it says

(8) Relative to each point P in supertime one point on my worldline is present, and it is the only point in all of spacetime that is present relative to P. (Furthermore, each point on my worldline is present at some point in supertime, and later points on my worldline are present relative to later points in supertime).

Figure 9.7 A solipsistic neo-classical theory.

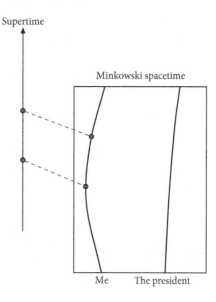

The "mismatch" that brings trouble to type-N theories is absent from (8). The structure of supertime may not fit well with the structure of the temporal aspects of Minkowski spacetime as a whole, but it looks exactly the same as the structure of the temporal aspects of my worldline.

Theory (8) is a solipsistic neo-classical theory. Figure 9.7 illustrates it. It does avoid the problems with type-N theories—but who could believe it? Who could believe that only his experiences were present?[18]

9.4 Superspacetime Theories

As unbelievable as they are, solipsistic neo-classical theories are on the right track. Once we take special relativity seriously it is a good idea to reject the claim that it is entire instants (in the sense of (D3)) that are present and say instead that it is individual spacetime points that are present. What believers in MST-Supertime should do is take this idea and go "fully relativistic." They should also reject classical supertime and replace it with (relativistic) "superspacetime." (Ultimately we need to know how to make MST-Supertense and MST-Time fully relativistic. But it is easiest to see how to do this by first seeing how to do it for MST-Supertime.)

Before saying anything about superspacetime I want to make some remarks about the claim that it is individual spacetime points, rather than entire instants, that are present. This proposal may seem to do some violence to the ordinary use of "present." But "present" is to some extent a technical term in the moving spotlight theory, and anyway, as Stein points out, there is an ordinary use of "present" to mean "here-now" ("On Relativity Theory and the Openness of the Future," p. 159). One uses it this way when, for example, one says "present" during roll-call. A more serious complaint is that a theory that says that it is individual spacetime points that are present departs too far from our ordinary ways of thinking. Callender presses this complaint; he says that the thesis that the present is a "spatially extended" region of spacetime is "part and parcel of our ordinary concept of time," and that there is no independent motivation for a theory that, like the one I am developing, says otherwise ("Shedding Light on

---

[18] Caspar Hare could. His book *On Myself, and Other, Less Important Subjects* is a defense of this claim. Whatever the merits of Hare's theory, (8) is worse. While Hare says that only his experiences are present, he goes on to say that from Skow's perspective Skow's experiences are present (and similarly for the rest of you). This is supposed to "soften the blow" of Hare's solipsism. But nothing like this is true in (8). (The best relativistic version of the moving spotlight theory will turn out to resemble Hare's version of solipsism.)

Time," p. S594). I think this is wrong. The independent motivation will be that the theory explains features of our experience as well as the classical theory.[19]

So what is superspacetime? The role of supertime in classical and neo-classical theories is to provide a set of perspectives on spacetime and its contents. In those theories, what spacetime and its contents are like—in particular, which region of spacetime is present—varies from perspective to perspective. Superspacetime plays the same role in the new theory. It is also a set of perspectives on spacetime and its contents, and which region (which point) of spacetime is present will vary from perspective to perspective. The difference is that in the fully relativistic moving spotlight theory the set of perspectives has the exact same geometry as spacetime itself. The theory restores the harmony between the geometry of spacetime and the geometry of the perspectives that went missing in the move from the classical theory to neo-classical theories.

In the classical theory the relationship between the regions that were present relative to two points in supertime matched the relationship between those two points themselves. Instants five seconds apart were present relative to points five superseconds apart. The same goes in the relativistic theory, except that the relevant relations are not those of order and distance but of order and interval. Here is what the theory says:

(9) Relative to each point in superspacetime exactly one spacetime point is present. Each spacetime point is present relative to some point in superspacetime. If P and Q are points of superspacetime, and the superinterval between P and Q is r, then the interval between the spacetime point that is present relative to P and the spacetime point that is present relative to Q is also r. Furthermore, if P is in the absolute future of Q, then the point that is present relative to P is in the absolute future of the point that is present relative to Q.

Figure 9.8 illustrates this theory.

So now we have a fully relativistic version of MST-Supertime—MST-Superspacetime we might call it. But the work of reconciling relativity and objective becoming is not done. The reasons for rejecting MST-Supertime

---

[19] Whether the classical theory explains those features particularly well is something I will address in Chapters 11 and 12. I briefly discuss relativistic theories in Section 12.1. See section V of Stein, "On Relativity Theory and the Openness of the Future," for an attempt to explain why, if spacetime is relativistic, our "ordinary concept of time" has it that the present is spatially extended. Callender is not the only philosopher to dislike a theory in which the present has vanishing spatial extent—though they are usually talking about presentism, not the moving spotlight theory. See for example Hinchliff, "A Defense of Presentism in a Relativistic Setting" (p. S579) and Saunders, "How Relativity Contradicts Presentism" (p. 286).

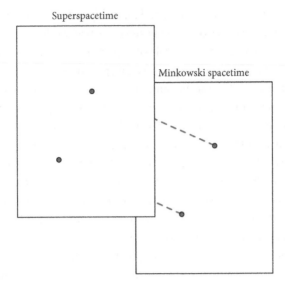

Figure 9.8 A superspacetime theory.

are also reasons for rejecting MST-Superspacetime. We need to see what the relativistic analogues of MST-Supertense and MST-Time look like. And I need to discuss some objections one might raise to these theories.

MST-Supertense does without supertime by using the metric supertense operators "It super-will be the case in r superseconds that . . ." and "It super-was the case r superseconds ago that . . .". A relativistic version of MST-Supertense that does without superspacetime obviously cannot use these same two operators. It makes no sense to ask which points in superspacetime are r superseconds to the superfuture of a given point in some absolute sense.

Since supertemporal separation in supertime has been replaced by superspace-time interval separation in superspacetime you might think that we should use instead the operators "It super-is the case superspacetime interval r away that . . ." But these will not work either.

The classical supertense operators are an adequate replacement for talk of supertime in part because everything we want to use supertime to say about how presentness moves through spacetime can be said using the supertense operators instead. But it is false that everything we want to use superspacetime to say about how presentness moves through Minkowski spacetime can be said using the operators "It super-is the case superspacetime interval r away that . . ." For example, we want to be able to say of each future spacetime point that it "will be" present, and we want to be able to say "how long it will be" until it is. We can say this in the language of MST-Superspacetime. Suppose that spacetime point Q is present at superspacetime point X, and that spacetime point P is in the absolute future of and at interval r away from Q. Then the way to say "when" P "will be" present is to say

(10) There is a point Y in superspacetime, in the absolute superfuture of and superinterval r away from X, such that P is present at Y.

But there is no way to say "when" P "will be" present using the operators "It super-is the case superspacetime interval r away that ..." The easiest way to see this is to add these operators to the language of MST-Superspacetime. For the sentences containing these operators that, in MST-Supertense, are true, are just the sentences containing them that, in MST-Superspacetime, are true at some given superspacetime point, which I shall take to be X. Now the only way to even try to say "when" P "will be" present is to say

(11) It super-is the case superinterval r away that P is present.[20]

But (11) is false at point X. For the truth-condition in MST-Superspacetime for (11) is

• "It super-is the case superspacetime interval r away that P is present" is true at superspacetime point X if and only if, at *every* superspacetime point superinterval r away from X, P is present.

The truth-condition must contain "every" because there are infinitely many superspacetime points superinterval r away from X. But P is present only at Y, not at any of the other points.

There is a way to formulate a relativistic version of MST-Supertense that avoids this problem. But no one should believe it. Fix a frame of reference F and use the following supertense operators:

• It super-will be the case in r superseconds relative to frame of reference F that ...
• It super-was the case r superseconds ago relative to F that ...
• It super-is the case r supermeters away in (superspatial) direction D relative to F that ...

These can do the work talk of superspacetime does because, given a point P in superspacetime, there is exactly one point Q that is (i) r superseconds to the superfuture of P relative to F, and (ii) r supermeters in direction D from P relative to F. So instead of saying what is true relative to each point in superspacetime

---

[20] Unlike in (10), there is nothing about the absolute future in (11). This could perhaps be fixed by using operators like "It super-is the case superinterval r toward the future that ..." This would not help with the problem I am about to describe.

we can fix a frame F and just say, for each r and direction D, which sentences A are such that both (i) it super-will be the case in r superseconds (or was the case r superseconds ago) relative to F that A is true, and (ii) it super-is the case r supermeters away in direction D relative to F that A is true. In this language MST-Supertense says

- Exactly one spacetime point N super-is present. For any positive r and s, and any direction D, X is a point in spacetime r seconds later-in-F (/earlier-in-F) than N and s meters away-in-direction-D-in-F from N if and only if both it super-will be (/super-was) the case in r superseconds relative to F that X is present, and it super-is the case s supermeters away in D relative to F that X is present.

This theory is unbelievable. First a small problem. The statement refers to a frame of reference F both as a frame in superspacetime and a frame in spacetime. The statement only does what it is supposed to if the theory ensures that these frames "line up."

But this just touches on the real problem. It was bad enough having "superseconds" appear in the fundamental vocabulary of the classical version of MST-Supertense. Here in the relativistic version we also have "supermeters." And what does "direction D" mean if there is no such thing as superspacetime for the directions to be directions in? And what does "relative to F" mean when there is no superspacetime for F to be a frame of reference in?[21]

Those are the objections to the relativistic version of MST-Supertense that lead me to dislike it. There are others that may move other people. Here is one. The theory has primitive sentential operators that are not tense operators. They are operators that stand to space the way that tense operators stand to time. But, the objection goes, the moving spotlight theory is supposed to be a theory in which there are deep differences between time and space. This theory appears to treat them the same.

I am not crazy about this objection, or this way of putting it anyway, because I do not think it is an independent constraint on the moving spotlight theory that it says that time and space are very different. What the moving spotlight theory must do is explain certain phenomena, especially phenomena about our

---

[21] There are relativistic versions of MST-Supertense that use different sets of primitive sentential operators. But the most obvious alternatives suffer from similar problems. For example, we could use operators like "It super-is the case superspacetime interval r away in superspatiotemporal direction D that . . ." But the reference here to superspatiotemporal directions is just as bad as the reference to superspatial directions. Another idea is to try to use non-metric tense operators. This can work when Newtonian spacetime is in the background—I described how in Section 7.2. But it seems to me hopeless in Minkowski spacetime.

experience of the passage of time, better than the block universe. If it does that then I do not see any further objection that time and space are too similar in the theory. (Whether it meets this explanatory demand is something that I have promised to get to later.)

So I do not think there are any good relativistic versions of MST-Supertense. But there is a good relativistic version of MST-Time. Probably we should call it MST-Spacetime. Where MST-Time identifies the perspectives on spatiotemporal reality with instants of time MST-Spacetime identifies them with points in space-time. The theory says that from the perspective of each point in spacetime that point and that point alone is present.

A natural question to ask about MST-Spacetime is, what about the past and the future? In the Newtonian spacetime version of MST-Time "spacetime point P is past" is defined to mean "P is earlier than (each of the points in) the present time." "P is future" has a similar definition. From the perspective of some time T, then, the present is the boundary between the past and the future. Figure 9.9 shows the familiar picture we get.

What definitions of "past" and "future" should we use in MST-Spacetime? I am not sure that there is—or that there needs to be—a unique answer. In addition, an answer that works well enough in Minkowski spacetime will almost certainly give weird results in some general relativistic spacetimes. Let's stick for now to Minkowski spacetime. Perhaps the most natural definitions are these:

- "P is past" means "P is inside the past light cone of the present point."
- "P is future" means "P is inside the future light cone of the present point."

Figure 9.9 Past, present, and future from the perspective of time T.

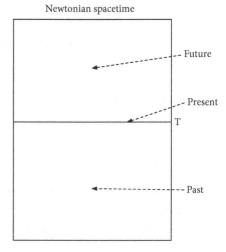

Newtonian spacetime

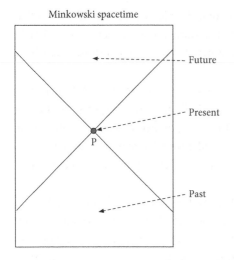

Minkowski spacetime

Figure 9.10 Past, present, and future from the perspective of P.

The picture we get with this theory is different from Figure 9.9. It looks like the picture in Figure 9.10. Two related differences are salient. First, from the perspective of P there are some points that are neither past, present, nor future. That never happens in the classical theory. Second, the present is not the boundary between the future and the past. In fact there is a four-dimensional region of spacetime separating the future from the past.

Are these differences problems for MST-Spacetime? They would be if it were a necessary truth that each point in spacetime is either future, present, or past. But I see no reason to think that this is a necessary truth. (One of the philosophical lessons special relativity teaches is to be wary of supposed necessary truths in this area.)

Another objection might be that *if there is objective becoming* then each point in spacetime is either future, present, or past. My response to this is similar to responses I have given to other objections. There are no independent reasons to require theories of objective becoming to say this. If the relativistic moving spotlight theory succeeds in explaining what it needs to explain without affirming it, then the theory is doing fine.

But can it explain what it needs to explain? Think about the objection in dynamic terms:[22] according to MST-Spacetime, the idea goes,

(12) Some future spacetime points will become past without ever being present!

One response is to say, so what? Our tensed language grew up with classical presuppositions. That (12) comes out true if special relativity is true might just show that our language has not caught up with scientific advances.

---

[22] This style of objection is originally due to Putnam ("Time and Physical Geometry," p. 246).

There is, however, some reason for proponents of the moving spotlight theory to be worried about (12). If (12) is true, it may be thought, then there can be people whose experiences are never present. And that looks like a return to solipsism—if not a complete return, at least a partial one.

This objection fails because (12) is in fact false in MST-Spacetime (from every perspective).[23] To see why we must first set out the theory's truth-conditions for tensed sentences of the form "it will be the case that P is future/present/past" and of the form "it is never the case that P is future/present/past." Back in Chapter 4 I set out weak and strong truth-conditions for tensed language in MST-Time. One relativistic version of the strong truth-conditions will say

(13) "It will be the case that P is future/present/past" is true from the perspective of X if and only if there is a point Y in the absolute future of X such that, from the perspective of Y, P is in the absolute future of / identical to / in the absolute past of the point that is present from the perspective of Y.

(14) "It is never the case that P is future/present/past" is true from the perspective of X if and only if there is no point in spacetime Y such that from the perspective of Y, P is in the absolute future of / identical to / in the absolute past of the point that is present from the perspective of Y.

Now fix a spacetime point A. Let B be any point that is in the future from the perspective of A. It is true that, from the perspective of A, B will be past. For B lies in the past light cone of a point that lies in the future light cone of A. But it is false that B will never be present. There is a spacetime point from the perspective of which B is present, namely B itself.[24]

If (12) seemed true to you it may have been because you reasoned as follows:[25]

Let P be the point of spacetime at which I am thinking these thoughts (see Figure 9.11). Then from the perspective of P these thoughts are present. Now let Q be a future

---

[23] A variant on the objection is: some past spacetime points never were present. But this is also false. (This is closer to how Hinchliff puts the objection in "A Defense of Presentism in a Relativistic Setting" (p. S579)—though, again, he is discussing relativistic versions of presentism—and is closer to Putnam's original formulation of the objection. See also Callender, "Shedding Light on Time" (p. S594).)

[24] Objection: "never" means "always-not," and "always" means "is, was, or will be." So (14) is wrong.

Reply: "always" is equivalent to "is, was, or will be" in classical contexts; but in classical contexts it is also equivalent (modulo reductionism about tense) to "at every spacetime point." Which equivalence should we focus on when figuring out what MST-Superspacetime says "always" means in Minkowski spacetime? I do not think we are required to focus on the first one.

Anyway, even if "always" does mean "is, was, or will be," (12) is still false. For since B is future from the perspective of A, "B will be present" is true from the perspective of A.

[25] This seems to be Putnam's and Callender's line of thought.

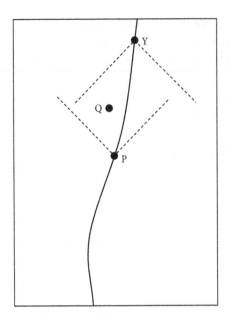

Figure 9.11 Q becomes past without ever being present?

spacetime point that does not lie on my worldline. There is no future point X on my worldline such that from the perspective of X Q is present. But there is a future point Y on my worldline such that from the perspective of Y Q is past. That is what I mean when I say that there are future points that become past without ever being present.

But so what? Why demand that points that are future and then past from perspectives I occupy are also present from a perspective I occupy? It should be enough that they are present from some perspective or other.[26]

My discussion of this objection does bring to the fore the solipsistic flavor of MST-Spacetime. If my experiences are present from some perspective, then from that perspective only my experiences are present. That might seem selfish. But the theory does not really say that I am special. For there are perspectives other than the ones I occupy, and from some of those perspectives other people's experiences are present.

The shift from a one-dimensional set of perspectives in neo-classical theories to a four-dimensional set in fully relativistic ones solves a lot of problems for the moving spotlight theory. But in one respect this shift is a liability. The moving spotlight theory is supposed to be a theory in which "the spotlight of presentness" moves through spacetime. How does the "motion of presentness" work in the

---

[26] Here is yet another version of the objection, pressed on me by Gordon Belot: if Q is at spacelike separation from P, then from the perspective of P, (i) it is not the case that Q is present; (ii) it is not the case that Q will be present; but (iii) Q will have been present. Here, again, I think moving spotlight theorists should go with the "so what?" response.

relativistic theory? Making sense of motion really does seem to require a one-dimensional set of (temporal) perspectives.

Even though the set of perspectives as a whole (that is, spacetime) is not one-dimensional, it has plenty of one-dimensional parts, namely timelike lines. Relative to a timelike line it is easy to make sense of the motion of presentness. From later perspectives on that line, later points (on that line) are present. Relative to a timelike line the spotlight moves toward the future along that very line. And the rate of time's passage in the relativistic theory comes out the way you expect. If talk of the spotlight's motion must be relativized to some timelike line then so must talk of the temporal distance the spotlight has traveled and the time it took the spotlight to make the trip. So if P and Q are two points on a timelike line L, then "the temporal distance the spotlight has traveled" is the temporal distance (relative to L) between the points that are present from the perspectives of P and Q (which, of course, are P and Q themselves). And the "time it took to make the trip" is the temporal distance (relative to L) between P and Q. So in the relativistic moving spotlight theory time also passes at one second per second.

The theory I have described owes a lot to Howard Stein's 1968 paper "On Einstein–Minkowski Space-Time." He proposed that "an event's [spacetime point's] present is constituted by itself alone" (p. 15). But it is unclear to me whether the theory Stein described is really MST-Spacetime.

There is one thing to be said in favor of Stein's theory being the same as MST-Spacetime. He presents it as a response to Putnam's argument, and Putnam certainly seemed to be arguing against objective becoming. Still, the whole Putnam–Stein dispute is slippery. Putnam, as I said, announces he will argue against the thesis

(15) All and only things that exist now are real.

That looks like presentism.[27] Putnam's argument then focuses on relations R such that

All and only things that bear R to me-now are real.

His argument proceeds by imposing restrictions on R. The central restrictions are that R must be transitive and definable using only relativistic notions. His argument against (15) then proceeds like this. Assume that I-now am not the only thing that is real. Some other thing W is also real, and W is in motion relative to me-now. Now the best candidate for R is the relation *x is simultaneous with y in*

---

[27] For now I count presentism as a theory of objective becoming.

y's frame of reference. W bears this relation to me-now. But there are things in my future light cone that bear this relation to W. So (15) is false.

But if presentism is Putnam's target this argument is doomed from the beginning. Putnam only considers one candidate for R. He does not show that every possible choice fails, and if his target is presentism there is an obvious choice that works. Just let R be the relation *x exists and y exists.* Presentists believe that all and only things that exist now bear this relation to me-now. And this relation is transitive and is definable using only logical notions, so is certainly definable using only relativistic notions. Putnam's argument only has a chance of working if we interpret it as directed at some other theory.[28]

That is what Stein does. Stein introduces the two-place predicate "X is real for Y" and assumes (i) that it applies to pairs of spacetime points, and (ii) it is not true by definition that it applies to every pair of spacetime points. He interprets Putnam as requiring that it have a definition in purely relativistic vocabulary. But what does this predicate mean? What is this notion we are trying to define? Putnam never used it and Stein does not tell us.

Stein is clearly uncomfortable with the slide from discussing what things are "real," real full stop, to what things are real for, or at, a given spacetime point. On page 19 he writes,

There is ... something disturbing in the statement of [Putnam's] principle: why is the notion "real" assumed, on the one hand, to be characterized by a relation R ... while it itself, on the other hand, is treated as a simple predicate?

Stein never answers this question. Later, on page 22, he writes,

My own view is that the issues of "reality" and "determinism" in the Rietdijk–Putnam form probably *cannot* be clarified.

It seems that Stein did not know what the theory he presents using "X is real for Y" really means, or whether it is intelligible at all. His main point seems to have been that it has the formal properties that Putnam said no theory using it could have.[29] For this reason some people come away from Stein's paper

---

[28] Actually, Putnam says that R must be definable using only relativistic notions and must be definable in a tenseless language (p. 241). The definition I have given on behalf of the presentist is in the present tense. But I do not think that this observation saves Putnam's argument. Some presentists will reject this additional requirement, since they reject reductionism about tense. Others may say that there is a tenseless form of the verb "to exist," and assert that everything that exists, exists, and vice versa. They can then use this tenseless form to define R.

[29] When Stein returned to this theory in "On Relativity Theory and the Openness of the Future" he made no further attempt to explain what "X is real for Y" means. Nor have I seen any of the philosophers working in the "Stein tradition" explain it. The purely formal way of thinking about his theory that Stein encouraged sometimes leads, I think, to strange places. In "The Definability of

skeptical that the theory he presents is a theory of objective becoming at all. Bourne is an example:[30]

Stein offers a definition of objective becoming . . . that is so deflated as to be indistinguishable from the tenseless [block universe] theory . . . how is his use of 'present' any different from the tenseless theorist's indexical use of 'present'? (*A Future for Presentism*, p. 165.)

The same question can be asked about MST-Time. I think the question can be answered, and I presented my answer back in Chapter 4. What I have done with Stein's theory is to add an interpretation that makes clear that it is a theory of objective becoming. The predicate "X is real for Y" means "X is present from the perspective of Y," and, as I discussed in Chapter 4, this "from the perspective of" talk can be understood so that the result is not just a version of reductionism about tense.

9.5 Beyond Special Relativity

Those who favor neo-classical spotlight theories might dismiss the conflict with special relativity because special relativity is false.[31] What happens when we consider "less false" physical theories? The answer is that things are no better for neo-classical spotlight theories and no worse for MST-Spacetime. MST-Spacetime is still the better theory.

The physics here is complicated but we only need a weak grasp of it to appreciate its bearing on our metaphysical questions. Special relativity has been superseded by general relativity. General relativity does not attribute, once and for all, a single geometry to spacetime. Instead there are many relativistic

Objective Becoming in Minkowski Spacetime" Clifton and Hogarth say that Stein's work is limited because it only discusses the predicate "(spacetime point) X is real for (spacetime point) Y"; they go on to consider theories that use the predicate "X is real for (inertial) observer O." But I do not think a believer in objective becoming should take seriously the suggestion that the perspectives on spacetime from which regions are present are individual people. The (robust) passage of time does not require the existence of people. (To be fair, Putnam writes in a way that encourages attention to this predicate. And the predicate Clifton and Hogarth end up investigating is actually "(spacetime point) X is real for (spacetime point) Y relative to timelike line L." This predicate can obviously be satisfied even if there are no people, though I think a theory using this predicate loses much of its appeal once it is made clear that in it regions are present not from the perspective of individual people, but merely from the perspective of pairs of points and lines.)

[30] Other examples include Le Poidevin, "Relative Realities" (p. 545) and Callender's review of Dorato's *Time and Reality* (pp. 118–19). Both are discussing Dorato's book, but the theory there is very similar to Stein's.

[31] A forceful dismissal may be found in Monton, "Presentism and Quantum Gravity," though (obviously) he is discussing presentism.

spacetimes, differing over where, and how much, they are curved.[32] The laws of general relativity relate the curvature of spacetime to the distribution of mass and energy in spacetime.

Generically speaking general relativity raises the same problems for neo-classical spotlight theories in which the present is spatially extended that special relativity does. A generic relativistic spacetime has many foliations of spacetime by instants, none of which is "geometrically special."

However, some relativistic spacetimes may raise new problems for neo-classical theories. There are relativistic spacetimes in which there are no foliations of spacetime by instants, and so no spatially extended regions that meet the minimal requirement for being a spatially extended present.[33]

But these spacetimes are not a problem for MST-Spacetime. This theory can continue to say that from the perspective of each spacetime point it and it alone is present.

Some think that other relativistic spacetimes solve old problems for neo-classical theories. One problem for type-N theories is that all frame-foliations are geometrically indiscernible. So if a given type-N theory says that present-ness moves along a given frame-foliation, it can say nothing about why it moves along that frame-foliation rather than another. But in some highly symmetric relativistic spacetimes there is one foliation by instants that is geometrically special. There are no other foliations geometrically indiscernible from it. The best general relativistic models of our universe are like this.[34] Also, some approaches to combining quantum mechanics with general relativity start by throwing out, as physically impossible, relativistic spacetimes that do not have a geometrically

---

[32] Curved relativistic spacetimes stand to Minkowski spacetime the way that curved surfaces in ordinary three-dimensional space stand to the Euclidean plane. In a small enough region a curved surface (the surface of a globe, or a lettuce leaf) looks almost exactly like the plane. Similarly, small enough regions of curved spacetime look almost exactly like Minkowski spacetime. (Differences in curvature are the most important, but not the only, way general relativistic spacetimes may differ. Two spacetimes may have the same curvature everywhere but differ topologically.)

[33] The main example of such a spacetime was discovered by Gödel before 1949. He described these results in "A Remark about the relationship between relativity theory and idealistic philosophy." Gödel assumed that if there is objective becoming then in every physically possible world there is a foliation of spacetime by instants. So, he argued, his spacetime shows that there is no objective becoming. Many who favor neo-classical theories just reject Gödel's assumption. Belot's "Dust, Time, and Symmetry" is a comprehensive discussion of Gödel's argument.

[34] Here I am thinking of Friedmann–Lemaître–Robertson–Walker (FLRW) spacetimes. The geometrically special foliation is special in several ways. It is the unique foliation into hypersurfaces of constant spatial curvature. It is also the unique foliation into hypersurfaces that are everywhere orthogonal to the worldlines of all material bodies. Belot's "Dust, Time, and Symmetry" contains a thorough discussion of the varieties of "absolute time" that may be found in general relativistic spacetimes.

special foliation by instants.[35] A neo-classical spotlight theory could say that it is the instants in the special foliation that are present.[36] If you think that the problem with type-N theories is that the foliation along which presentness moves is not geometrically special then these relativistic spacetimes are the answer to your prayers.[37]

Another road to this kind of neo-classical theory ignores general relativity and just focuses on what it takes to reconcile quantum mechanics with special relativity. (Since gravitational effects are being ignored we are still dealing with false physics—though perhaps "less false.") Some theories do this by adding a notion of simultaneity the way that type-G theories do.[38]

It still seems to me that MST-Spacetime is better than these neo-classical theories. Go back to my discussion of deviant versions of the moving spotlight theory in Newtonian spacetime. They say that the regions that are present form a foliation of spacetime, but that the leaves of the foliation are not (Newtonian) instants. I said that these theories are false because they permit an extended segment of someone's worldline to be present at a single point in supertime.

I think that this kind of argument is the only basis on which a believer in objective becoming can legitimately reject a foliation of spacetime by instants. In Newtonian spacetime this kind of argument selects a unique foliation to be the set of regions that are, successively, present. But it does not select a unique foliation in the relativistic spacetimes that have geometrically special foliations. There are other foliations of these spacetimes by instants. So what if these other foliations are not geometrically special? It is still impossible for an extended segment of an observer's worldline to lie entirely inside one of them. Even if presentness moved along the instants in one of the non-special foliations each person's experiences would be successively present, with no two experiences being present

---

[35] The "constant mean curvature (CMC) approach to quantum gravity" is the main example.

[36] The collection *Einstein, Relativity and Absolute Simultaneity* (Craig and Smith, eds.) contains several recent papers that defend objective becoming (or presentism) using one or another of these ideas. Jeans, in "Man and Universe," was one of the first to say that FLRW spacetimes solve the problem Minkowski spacetime raises for objective becoming. Monton, in "Presentism and Quantum Gravity," was one of the first to discuss the CMC approach to quantum gravity when defending presentism. Wüthrich, in "No Presentism in Quantum Gravity," complains that the CMC approach is so far away from being an adequate physical theory that it is unwise to appeal to it when doing metaphysics.

[37] This problem with type-N theories came up earlier when I discussed their conflict with principle (4).

[38] I am thinking of some versions of Bohmian mechanics and GRW. See Callender, "On Finding 'Real' Time in Quantum Mechanics," for more details on these theories and on why there is pressure within quantum mechanics to add a notion of simultaneity to Minkowski spacetime.

at a single point in supertime. So these theories can do the explanatory work that the moving spotlight theory needs to do.[39]

9.6 Appendix: Spatially Extended Presents without Underdetermination

One of my arguments for taking the regions that are present to be individual points is that neo-classical theories with spatially extended presents face an underdetermination problem. The weak point of this argument is that there are moving spotlight theories with spatially extended presents that are not neo-classical.

One brute force approach to building a theory like that is to modify MST-Spacetime as follows. Instead of identifying the perspectives on spacetime with individual spacetime points, identify them with timelike (straight) lines, each with a distinguished point. From the perspective of (L,P) the region of spacetime that is present is the spacelike line orthogonal to L through P.

This theory has its problems, one of which is that each region that is present is present from many perspectives. A better approach, suggested to me by Gordon Belot, is to identify the perspectives with the instants of time (in the sense of D3). Then from the perspective of instant T it is instant T itself that is present. This relativistic theory is in some ways the minimal departure from MST-Time, differing only in using a relativistic definition of "instant." It is not a neo-classical theory, since the set of instants, and so the set of perspectives, in Minkowski spacetime is not linearly ordered.

Is this theory better than MST-Spacetime? One worry is that, in this new theory, the experiences I am now having are present from the perspectives of many instants. Didn't I earlier treat this as a decisive objection to another theory? Yes, but in this case there is a difference. Although there are distinct instants from the perspective of which these experiences are present, neither of those instants is earlier than, or later than, the other. Maybe fans of objective becoming are happy as long as the experiences I am now having are not present from the perspective of any later instant.

And this theory does have an advantage over MST-Spacetime. It is hard to see how a spotlight that shines on just one spacetime point can explain anything

---

[39] Callender, in "On Finding 'Real' Time in Quantum Mechanics," and Wüthrich, in "The Fate of Presentism in Modern Physics," also question why there should be a connection between being the geometrically preferred foliation of spacetime and being the foliation along which presentness moves, though not on these grounds.

about my experience. I have followed common practice by assuming that our spatiotemporal careers may be represented by timelike lines. But really we extend in all three spatial dimensions; our careers must be represented by "worldtubes," not worldlines. Now we want the theory to say something like: the experience I am having is the one that is present. But even if the spacetime point that is present falls inside my worldtube, it is far from clear that there is one unique experience of mine that the spotlight is shining on. In the theory just described, by contrast, the spotlight shines on a three-dimensional submanifold, just as in the classical theory. So the new theory is no worse off than the classical theory when it comes to explaining anything about experience. If the intersection of my worldtube with an instant suffices to single out a unique experience in the classical theory, it does so as well in the relativistic one.

We should remember, though, that moving spotlight theories with spatially extended presents, neo-classical or not, face a problem that MST-Spacetime avoids, namely the existence of general relativistic worlds in which there are no instants (in the sense of D3). A fan of spatially extended presents could just throw these worlds out, saying that even if the usual laws of general relativity say those worlds are possible, the laws of metaphysics say they are not. A move like this is not uncommon, though I would prefer to avoid it.

9.7 Appendix: Relativity and the Block Universe

The classical block universe theory, like the classical moving spotlight theory, is inconsistent with special relativity. The theory gives this tenseless and A-vocabulary free truth-condition for a typical past-tensed sentence:

- "X was F" is true at T if and only if X is F at a time earlier than T.

But if "time" is defined by (D1) this makes no sense, or gets truth-values badly wrong. What should replace it?

Certainly truth-values of tensed sentences must be relativized to spacetime points, not to instants of time. But that is not the only change that must be made. We need to say what "earlier than" means. There are a couple of things it could mean. Here is one truth-condition our typical past-tensed sentence could have. This truth-condition relativizes truth to points of spacetime and frames of reference:

- "X was G" is true at spacetime point P and frame of reference F if and only if X is G at some spacetime point earlier-in-F than P.

The idea is that whenever someone says something of the form "X was G" they say it at some spacetime point P, and their worldline at P determines a frame F (F is the frame determined by the straight timelike line tangent to their worldline at P). This point and frame are to be used when evaluating the utterance.

There is some idealization going on here. Point-particles, things that have zero spatial extent, have worldlines, but larger things like human beings have world-tubes. And there is no such thing as the line tangent to a worldtube at a point. The more general point here is that an individual's worldtube plus a point inside that worldtube will not determine a unique frame of reference. But this obser-vation does not make the truth-condition just stated completely useless. There was already some idealization in the classical theory's talk of "the" time at which a sentence is used. Usually that idealization is harmless. We get the right results no matter which instant during the stretch of time required to speak the sen-tence is chosen as the time of use. Similarly here a worldtube and a point inside it (which we regard as the point of speech) will determine a range of candidate frames of reference (say, the frames containing that point in which at least one of the speaker's parts is at rest), and we can hope that we get the right results no matter which of these frames we use to evaluate the utterance.

The second truth-condition for tensed talk relativizes truth just to a point of spacetime:

- "X was G" is true at spacetime point P if and only if X is G at some spacetime point absolutely earlier than P.

Most things we say in ordinary life will get the same truth-value on either of these truth-conditions.

The classical block universe's treatment of A-vocabulary also needs to be refor-mulated in light of special relativity. "Present" is sometimes used as a predicate of spacetime points and events. In the classical theory predicates have extensions only relative to times, just as sentences have truth-values only relative to times. The theory says that the extension of "present" relative to T contains the space-time points in, and events that occur at, T. To make this relativistic we again need to replace relativity to times with relativity to points of spacetime. Then there are analogues of the proposals for tensed sentences. We could say that the extension of "present" is relative to a point P and a frame F, that it is the set of points and events simultaneous-in-F with P. Or we could say that the extension of "present" is relative just to points. Here there are many options. We could say that relative to P the extension contains only P. Or we could say that it contains P together with all points at negative intervals from it. A fourth option, due to Gibson and Pooley in "Relativistic Persistence," is to say that the extension of "present" at P is a disk, thin in timelike directions and wide in spacelike directions, centered on P.[40] Or maybe we should not choose once and for all. Maybe which of these accounts is correct varies from context to context. Or maybe there is no fact of the matter about which is correct. Believers in the block universe theory should not regard these as deep philosophical questions.

Whatever truth-conditions, or range of candidate truth-conditions, we adopt, there are bound to be circumstances in which strange-sounding sentences come out true. Just consider again the sentence from footnote 26: "Q will have been present, but it is not the case that Q is or will be present," uttered at a point P at spacelike separation from Q. This example is set in Minkowski spacetime; things are bound to go even more haywire if we consider the whole range of general relativistic spacetimes. Probably we block universe theorists should say that, given the truth of relativity, tensed language is seriously defective.

[40] Their proposal makes more sense if we remove the simplification of thinking that the extension of "present" is relative to a point of spacetime. Really it is relative to a context. A context determines a point in spacetime as the point of speech of the context, but contains other information as well. The dimensions of the disk will depend on other features of the context, including the rate of the biological processes of the speaker, especially the processes inside his brain responsible for perception.

10

Can We Move through Time?

10.1 Making Sense of the Idea

For six chapters I have been discussing the idea that the passage of time consists in time itself moving, or at least changing—"time flows like a river." But there is another idea about what robust passage might be, one that has been on hold since Chapter 3. It is the idea that the passage of time consists in our moving through time—"we are sailors on the ocean of time." Instead of a spotlight passing over spacetime and its contents we have the contents themselves "marching upward" through spacetime. I have been arguing that the first idea is not inconsistent or easily refuted. But what about the second idea? Can any sense be made of it?

Motion requires three things: the thing that moves, the dimension or "space" in which it moves, and the dimension with respect to which it moves. If I am moving through time then, clearly, the thing moving is me and the dimension in which I move is time. But what is the dimension with respect to which I move? Michael Lockwood held that since there is no dimension available motion through time is impossible:

When shown a space-time diagram . . . incorporating a world-line that is intended to represent the life of a human being, one has an almost irresistible urge to interpret it dynamically. One tends to think of the individual as like an ant that crawls along the line from one end to the other. . . . But, on the face of it, no such interpretation is logically permissible. *For time is already included in the diagram.* (*Mind, Brain, and the Quantum*, p. 261.)

But we have been over this ground before. The same problem came up when I discussed what it takes for time itself to move. The preliminary solution there was to say that the dimension with respect to which "presentness moves" is supertime. The analogous solution here is to let supertime be the dimension with respect to which we move through time. The theory that results looks like this:

In addition to spacetime there is also supertime. At each point in supertime each material thing is located at exactly one time, and it is the same time for each material thing. In four-dimensional terms, relative to each point in supertime each material thing is located at some region of spacetime that sits entirely inside a "time-slice" of spacetime. Relative to later points in supertime material things are located at later times, in such a way that they move at a constant rate through time.[1]

Figure 10.1 illustrates the theory. The left panel shows my spatiotemporal location relative to supertime point P1; the right shows my location relative to a later point P2. For comparison Figure 10.2 illustrates the block universe theory. (In both figures I move through space, by moving to the left.)

Interestingly, after the passage quoted Lockwood goes on to write that "It isn't as though the situation were to be described by a continuous succession of space-time diagrams . . . in which the 'ant' is progressively further along the world-line" (p. 262). That is exactly how the situation is described in this theory.

These two pictures and theories differ over how they treat my spatiotemporal location. In the block universe theory my spatiotemporal location is absolute. There is a four-dimensional region of spacetime R such that I am "just plain" located at R; I am not located at R relative to this or that. In the new theory of objective becoming, by contrast, my spatiotemporal location is relative, to points of supertime. And I am always—at every point in supertime—located at a three-dimensional region of spacetime.

I do not mean to suggest that there is no way to make sense of talk of absolute spatiotemporal locations in the new theory. Here is one way to do it. Let S be the set of regions of spacetime that I am located at relative to some point of supertime or other. Let R be the fusion of S (the smallest region containing all of the regions

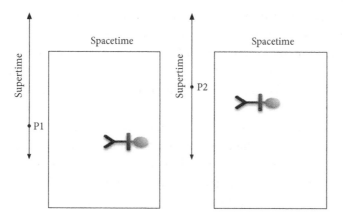

Figure 10.1 Me, moving through time.

---

[1] Zimmerman discusses a theory like this, which he regards as a version of presentism, in "Presentism and the Space-Time Manifold."

Me in spacetime   An outline of my spacetime location

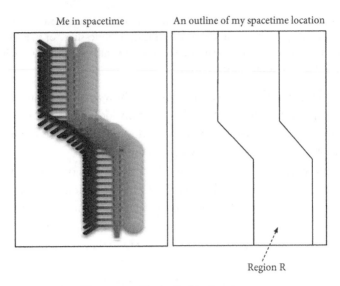

Region R

Figure 10.2 Me, in the block universe.

in S). Then we could call R my absolute spatiotemporal location. But I think this would be misleading. The fundamental location claims in the new theory are the relative ones, the claims about where in spacetime I am at each point in supertime. The absolute claims are derived from these relative claims. Since we are interested in what these theories say the world is like fundamentally speaking, we should ignore the non-fundamental claims about absolute location.

Just as we eliminated supertime from the moving spotlight theory, we will want to eliminate supertime from this new theory. The same two strategies are available. We can eliminate supertime in favor of primitive supertense operators, or eliminate supertime in favor of time and talk of what is true "from the perspective of" each time. Having said this, I will stick for the most part to the supertime version of the theory.

The new theory says that we literally move through time. It is supposed to be a theory that has what the block universe lacks, namely robust passage of time. So do I not move through time in the block universe? It does not seem so. Motion through time requires variation in my location in time. But there does not seem to be any such variation in the block universe theory. There I am, located in region R, once and for all.

However, just as there is a way to define "absolute spatiotemporal location" in the new theory, there is a way to define "spatiotemporal location relative to time T" in the block universe theory. It can be defined like this:

(D1) X's spatiotemporal location relative to T is R $=_{df}$ R is the largest part of X's absolute location that is entirely confined to T.

If we use this notion then we can say, even in the block universe theory, that there is variation in my temporal location. Relative to a time T1 I am located at T1 (at a region of spacetime that lies entirely within T1); but relative to a later time T2 I am not, I am instead located at T2. So do I move through time in the block universe after all?

No. Genuine motion requires at least the conceptual possibility of rest. Trains move, but they do not, as far as the meanings of our words go, need to move.[2] Perhaps some things do move of necessity, but that necessity is never conceptual. It is necessary that photons move. Under no relativistically possible conditions is there any frame of reference in which a photon is at rest. But this necessity is physical. It is not conceptually necessary that photons move.

If, however, definition (D1) permits us to say that I move through time, then it also permits us to say that it is necessary that I move through time (given that I persist through time at all, that I am not instantaneous). It is impossible that I stay put. And this necessity and impossibility comes straight from definition (D1); it is a conceptual necessity and impossibility. So this is not genuine motion through time.

10.2 Moving through Time and Persisting through Time

The contrast between the new theory of objective becoming and the block universe resembles, and is connected to, a contrast between two theories of persistence through time. One of those theories, the endurance theory, fits better with the new theory than it does with the block universe. This is a relatively weak claim. I am not saying that the endurance theory is inconsistent with the block universe. But there are challenges to making it intelligible in the block universe theory that do not arise in the new theory.

You, me, pianos, and trombones all persist through time. That is, we presently exist, and we will continue to exist. Moreover, we will change. My hair will be gray and my piano will be out of tune. The perdurance theory of persistence starts with the block universe's "initial" truth-conditions for *de re*-tensed predications

---

[2] This is right even if we reject Newtonian spacetime and so reject absolute motion. Then something is moving or at rest only relative to some frame of reference. My claim becomes the claim that motion-in-frame-F requires the conceptual possibility of rest-in-frame-F.

and subjects them to further analysis. For example, the initial analysis of simple "future-tensed" statements is

- "X will be F" is true at T if and only if there is a time U later than T such that X is F at U.

The perdurance theory uses temporal parts to analyze the right-hand side:

- X is F at T if and only if there is an instantaneous temporal part of X that exists at T and is (without temporal qualification) F.

We started with *de re*-tensed predications; the block universe theory analyzes them in terms of tenseless but "time-indexed" predication; the perdurance theory goes on to analyze tenseless time-indexed predication in terms of temporal parts and index-free predication. (Unless they deny that things persist, those who accept the perdurance theory also accept the "doctrine of temporal parts," the thesis that each temporal thing has an instantaneous temporal part at each time it exists.)

The perdurance theory treats time-indexed predication the way we usually think place-indexed predication works. We usually think that for something to be red here is for it to have a part here that is simply red; the perdurance theory says that to be red later is to have a later part that is simply red. An alternative theory is the endurance theory. Saying just what the endurance theory is a tricky business. We can certainly say this much, though: endurance theorists who also accept the block universe theory of time refuse to say anything informative about the truth-conditions for sentences that express tenseless time-indexed predication (sentences of the form "X is F at T"). Tenseless time-indexed predication, they say, is fundamental. (If things do not have temporal parts then, clearly, time-indexed predication cannot be analyzed using temporal parts and index-free predication. But even endurantists who accept temporal parts reject this analysis. I should note that the endurance theory and the perdurance theory are not the only two theories of persistence; I will discuss another, the stage theory, in Chapter 12.)

Now time and space are much more similar in the block universe theory than they are in theories with robust passage. Since the perdurance theory analyzes time-indexed predication in the same way we ordinarily think spatial-location-indexed predication is analyzed, it fits into the block universe very well.

I said that the perdurance theory fits well into the block universe theory. How well does it fit with this new theory of objective becoming? They are inconsistent.

The perdurance theory as I formulated it presupposes reductionism about tense, which the new theory rejects.[3]

The situation is different with the endurance theory. The rhetoric surrounding the theory fits better with the new theory of objective becoming. In a way, that theory is a more natural home for the endurance theory.

Why? One piece of rhetoric that accompanies the endurance theory is talk of material things "sweeping through spacetime." This is supposed to contrast with talk of material things being "spread out in spacetime," talk that is supposed to be more apt in the perdurance theory. But sweeping through spacetime seems like a way of moving through spacetime. And it is in the new theory, not the block universe, that things move through spacetime.[4]

Another piece of rhetoric that accompanies the endurance theory is talk of persisting things being "wholly present" at each time they exist. This is supposed to mark that the theory treats persistence through time differently from extension in space, for spatially extended things are only partially present at each place they occupy (they are partially present at a place by having a part that is wholly present). This claim, that persisting things are wholly present at each time, is mysterious and in need of clarification. It has received several clarifications, but I want to focus here just on one. Some endurantists interpret "X is wholly present at T" as a claim about X's location.[5] They assume that the block universe theory is true and formulate the endurance theory like this:

(E) A persisting thing has many spatiotemporal locations. That is, there are many regions of spacetime at which it is located. Each of these regions is instantaneous (has zero temporal extent). If the persisting thing exists at

---

[3] The doctrine of temporal parts, however, is consistent with the new theory. (A statement of the doctrine that is consistent with the supertensed version of the theory will resemble the statement of it that is consistent with presentism, for which see Sider, *Four-Dimensionalism*, chapter 3.) Why would anyone want to combine the doctrine of temporal parts with this new theory of objective becoming? What good are temporal parts if you are not going to use them to analyze tenseless time-indexed predication? There are other reasons to believe in them. Suppose that the new theory of objective becoming is true and that a clay statue exists at a certain time. Suppose further that the clay super-is about to be squashed into a lump. Then, one might think, the clay but not the statue super-will continue to exist. If this is right the statue and the clay are distinct even though they coincide. A standard motivation for the doctrine of temporal parts is that it draws the sting out of this claim by allowing one to say that the statue is a temporal part of the clay.

[4] Sider repeatedly uses "sweep through spacetime" when presenting the endurance theory; see pages 67 and 69 of *Four-Dimensionalism*. Hawley skips past the word "sweep" and directly attributes to endurance theorists the idea that "objects seem to 'move' through time in their entirety" (*How Things Persist*, p. 10).

[5] See especially Cody Gilmore's "Where in the Relativistic World Are We?" and "Persistence and Location in Relativistic Spacetime." On another interpretation "X is wholly present at T" is a claim about X's parts, not about its location. I will say something about that interpretation in the appendix.

time T then among its locations is exactly one instantaneous region that is part of the "T-time-slice" of spacetime. And these are all of its locations.

Claim (E) is supposed to contrast with the perdurance theory. But the perdurance theory as I stated it says nothing about how many spatiotemporal locations a persisting thing has, or what those locations are like. Those who endorse (E) expect the perdurance theory to say this:

(P) A persisting thing has exactly one spatiotemporal location. A human being, for example, is located at a single four-dimensional region of spacetime, like region R in Figure 10.2.

Claims (E) and (P) sound a lot like the characterizations I gave above of the new theory of objective becoming and the block universe theory. If we construe (E) as using the notion of relative location that appears in the new theory and (P) as using the notion of absolute location that appears in the block universe theory then they look like this:

(E-rev) A persisting thing X has many relative spatiotemporal locations. There are many regions of spacetime R with the property that X is located at R relative to some supertime or other. For a human being these regions are instantaneous and extend in all three spatial dimensions.

(P-rev) A persisting thing X has exactly one absolute spatiotemporal location. For a human being this region is four-dimensional.

So the endurance theory's rhetoric of "wholly present" fits well with the new theory because if that theory is true then (E) is naturally interpreted as (E-rev), a claim that actually follows from the theory.

Many endurance theorists intend (E) to be consistent with the block universe theory. They do not want to interpret it as (E-rev). But what then does it mean, and how is the debate over which of (E) or (P) is true to be understood?

One obstacle to understanding that debate is that "absolute location" has many possible meanings. For the next few paragraphs I will be talking about the notion of absolute location that figures in the block universe theory, so for simplicity I will drop the "absolute" and just use "location." To get a sense of the many meanings "location" may have ignore time and think just about spatial location. Certainly there is some sense of "location" according to which I have many spatial locations. Take some point of space P one foot above the chair I am currently sitting in. I am certainly located, in some sense, at P. But, one wants to say, I am not "exactly located" at P. I am exactly located at exactly one region of space H, one that is human-shaped. (Point P, of course, falls inside region H.) In the first sense of "location" it is a definitional truth that if X is located at R and S is a subregion of R then X is located at S. In the second sense, the one I am using "exact location" for, this is not even true, much less a definitional truth.

Now go back to the spacetime point of view. One might worry that, interpreted so as to be compatible with the block universe theory, (E) and (P) do not really conflict because they use "location" in different senses. Maybe "X is located at R" in (P) means "X is exactly located at R" while in (E) it means something like "R is an instantaneous part of X's exact location."

Those who accept (E) insist that the parties to this debate are not talking past each other, or at least that they need not be. (E) does conflict with (P) provided that both are understood so that "location" means "exact location."

But this is puzzling. Isn't it part of the meaning of "exact location" that, assuming the block universe theory, each persisting thing has exactly one exact location in spacetime? If so then there is no room for endurantists and perdurantists to disagree about how many exact locations a persisting thing has.[6]

Some endurantists say that while there may be a meaning for "exact location" that makes "If the block universe theory is true, then each thing has only one exact location" a definitional truth, there is another meaning it may have, one that leaves it open how many exact locations something has. If we understand (E) and (P) so that "location" has this second meaning, they say, then neither thesis will be false by definition.

The dispute now moves up a level. Before we can investigate how many exact spatiotemporal locations a persisting thing has we need to investigate what meanings "exact location" may have. Some endurantists say two, some perdurantists say just one, the one that makes (E) false by definition.

Those who say that "exact location" has just this one meaning have an argument for their view. Take a meaning that everyone agrees "location" may have: the meaning that makes "I am located at a spacetime point P that is part of an instant of time in 2012 and is one foot above this chair" true and makes "I am located at some region of spacetime that extends from 1900 to 2100 and is two miles in (spatial) diameter" false. Let us use "location(1)" for this sense of "location."[7] The following is a definitional truth:

- If X is located(1) at R and X is located(1) at S then X is located(1) at R + S.

---

[6] I tacitly assumed this view when I talked about the pictures in Figure 10.2. I said that according to the block universe theory locations are absolute, and that I am located at R, a certain four-dimensional region of spacetime. And I meant this to be true whether it is the perdurance or the endurance theory that is correct.

[7] There is another meaning for "location," call it "location(2)," that makes the second sentence true and the first false, but it will play no role in what follows. If we had already settled what "exact

It follows from this that there is a unique, largest region at which something is located(1). So one definition of "X is exactly located at R" is

(D2)   X is exactly located at R $=_{df}$ R is the largest region at which X is located(1).

Claim (E) is automatically false if in it "location" means "exact location" as defined in (D2). But there is no other definition of "exact location" in terms of "location(1)."

This argument does not show that there is no meaning for "exact location" on which (E) is not false by definition. It just shows that we cannot get at that meaning by writing a definition using "location(1)."

Still, I think the argument adds to the case that the new theory of objective becoming is the natural home for the endurance theory as characterized by (E). We have already seen that it is easy to understand (E) so that it is unproblematically true in the new theory. Is there a way to understand (E) so that it is compatible with the block universe theory? The best way to argue for a yes answer—define the relevant sense of "location" in uncontroversial terms—does not work. Some endurantists remain convinced that answer is yes. What makes them so confident? I wonder about the extent to which the theory in which we move through time is lurking in the background, making it seem obvious that (E) is, if not true, at least consistent; but lurking invisibly, so that it is not noticed that the obvious consistency of (E) presupposes a theory of time incompatible with the block universe.

10.3 Evaluating the Theory

How does the new theory of objective becoming stack up against the moving spotlight theory? In some respects it is just as good. The challenge to say how fast time passes is as easily met in the one theory as in the other. In the new theory this question becomes, how fast are we moving through time? And it is a substantive part of the theory that if P is one supersecond later than Q then my location relative to P is one second later than my location relative to Q.

In other respects the new theory might seem the better theory. The standard version of MST-Supertime says that the only thing that changes from one point

---

location" means we could define these others in terms of it: we could define "X is located(1) at R" as "R is part of X's exact location" and "X is located(2) at R" as "X's exact location is part of R." But what "exact location" may mean is what is in question (note that these definitions use the sense of "exact location" on which exact locations are unique). The philosopher who has done most to identify the relations among the senses of "location" is Josh Parsons. The argument I am summarizing may be found in his paper "Theories of Location."

in supertime to another is which time is present. This change does not connect with our experiences, or anything we might care about. In the new theory, by contrast, lots of things change from one point of supertime to another. There is supertemporal variation in my spatiotemporal location, and also in my size, shape, in whether I am in pain, and so on.

The new theory lacks this advantage over "non-standard" versions of the moving spotlight theory, in which more changes from supertime to supertime than which time is present. I will describe a theory like that in Chapter 12. Really the theory will be a hybrid of the moving spotlight and the new theory in which we move through time.

But for now let me stick to discussing pure versions of the theories. I have mentioned some respects in which the new theory is better than the moving spotlight theory. There are also respects in which it is worse. Each theory attempts to capture a different picture of what the passage of time consists in. The moving spotlight theory does a pretty good job capturing the picture of "time flowing like a river." But the new theory does a less good job capturing the picture of us "sailing on the ocean of time." For it is supposed to be part of that picture that events float on the ocean of time and we come across them as we sail. But in the theory there are no future events. Relative to a point in supertime P, there is a single time T that all material things inhabit. All times later than T are empty. So as we "sail into the future" we are heading into an empty ocean. Of course, things super-will happen. Relative to a point in supertime later than P, a time later than T will have things and events in it. But they are not "already there" waiting for us.

A more serious problem for the theory is how well it meets the challenge from relativity. The spacetime version of the moving spotlight theory says that I exist, spread out in four dimensions, but the spotlight shines on at most one of the spacetime points I occupy. The spacetime version of the new theory will say that, from any single perspective, all of spacetime is empty save for just one spacetime point. Where am I in this theory? I can see how I may fit into a three-dimensional time-slice of spacetime, as the classical moving us theory says. But I do not see how I can fit into a dimensionless point. So the only way to make the new theory relativistic is to follow one of the strategies for making a relativistic moving spotlight theory with spatially extended presents. One strategy was to use neo-classical supertime; I argued that that strategy was a bad one. Following the strategy from the first appendix to Chapter 9 is more promising.

Still, all things considered I think that the moving spotlight theories, and the hybrid theory I will describe, are the better theories. For the remainder of the book I will focus mostly on them.

11

Passage and Experience, I

11.1 Arguing for Robust Passage

The fact that the moving spotlight theory is internally consistent, and compatible with the theory of relativity, does not alone make it worthy of belief. To command belief there must be some good arguments in its favor.

You might say that it is just obvious that time passes, so if there are no obstacles to believing the moving spotlight theory we should believe it. But this only follows if it is just obvious that there is *robust* passage. Is that really obvious? Not to me.

The usual instruction given to someone looking for evidence of robust passage is: attend to your experience! The evidence is *right there*. Schlesinger for example wrote,

there is hardly any experience that seems more persistently and immediately given to us than the relentless flow of time. ("The Stream of Time," p. 257.)

But this way of describing the evidence our experience provides raises more questions than it answers. What is it for something to be "immediately given" to me in experience? This is a technical philosophical term that requires explanation. Different explanations of it lead to two different kinds of argument "from experience" for the existence of robust passage. Neither of them has any hope of getting off the ground. But there is another argument from experience, not connected to the idea that passage is immediately given in experience, that is better, or at least more interesting. I call it the argument from the presented experience, and spend Chapter 12 discussing it.

11.2 The Argument from the Content of Experience, I

Some defenders of the block universe theory have thought they could defuse all arguments from experience without having to figure out just how those arguments work. They argue: supposing that the moving spotlight theory is true, my experience would be just the same if the block universe theory were true, and vice versa. So experience cannot favor the moving spotlight theory over the block universe theory.[1]

This is dangerously close to one kind of argument for skepticism about the external world. Supposing that I have hands, my experience would be just the same if I were a handless brain in a vat hooked up to a computer running the right kind of program. So my experience's being as they are cannot favor the hypothesis that I have hands over the hypothesis that I do not. There are various diagnoses of the flaws in this argument for skepticism. One is particularly relevant here. Even if it is true that each of two hypotheses is consistent with my experience being as it is, it does not follow that the fact that my experience is that way fails to favor one of the hypotheses over the other. In general, a body of evidence can favor one hypothesis over another without entailing the first and being inconsistent with the second.

Some defenders of the block universe theory might retrench at this point. They might say that when the evidence is consistent with two theories it favors the simpler theory. And, they might continue, the block universe theory is certainly simpler than the moving spotlight theory. I, however, do not think that "a body of evidence favors the simplest theory consistent with it" is a basic epistemological principle. The more basic principle says that the evidence favors the theory that best explains it. Sometimes that is the simpler theory, but not necessarily. And if it is explanation that matters then it is not obvious ahead of time that the block universe theory will come out ahead.

One way to put together an argument from experience, then, is to say that the moving spotlight theory better explains some feature of our experience. Here is a line of thought that leads to an argument like this. Many who accept the block universe theory admit that there seems to be robust passage of time. They then go on to say that this is just an illusion. Smart, for example, begins "Time and Becoming" with "I intend to argue . . . that the alleged passage of time or pure becoming is an illusion" (p. 3). And Falk, just to have a second example, says that his version of the block universe theory does "not treat the whoosh and whiz [that is, the passage of time] as illusion merely," thereby admitting that it does treat it as an illusion.[2] Moving spotlight theorists, on the other hand, say that things are as they seem. There really is robust passage.

A moving spotlight theorist might stop here and argue that his theory is superior because it is only in his theory that things are as they seem. But we often

---

[1] Price gives this argument in *Time's Arrow and Archimedes' Point* (pp. 14–15) and Prosser endorses it in "Passage and Perception." Maudlin denies the premise in "On the Passing of Time." But the argument is bad even if the premise is granted.
[2] "Time Plus the Whoosh and Whiz," p. 215.

have excellent evidence that things are not as they seem. A stick in a cup of water may look bent, but if we have examined it earlier we may know that it is not. A believer in the block universe might also have an excellent story to tell about why things are not as they seem.

So we need to look at what each theory says about why there seems to be robust passage. The argument will be that the moving spotlight theory's explanation is better than the explanations believers in the block universe can come up with, and so, other things being equal, we should believe the moving spotlight theory.

First we should dwell for a minute on the claim that, if the block universe theory is true, then while there seems to be robust passage, this is just an illusion. Think about an ordinary illusion, say the Müller-Lyer illusion. This illusion consists in two lines that are the same length, but nevertheless look to be different lengths (Figure 11.1). In general, a visual illusion is a scene in which things are not the way they look. Not all illusions are visual illusions, of course. But it is not hard to characterize illusions without mentioning a particular sensory modality. A perceptual illusion is a situation in which things "perceptually seem" to be a certain way, but are not.

Now to say that the two lines in the Müller-Lyer illusion look to be different lengths is to say that they are visually represented as having different lengths. It is to say that the visual experiences I have when I look at them are accurate, or veridical, only if the lines are different lengths. If we indulge in some proposition-talk then the claim is that my experience when I look at the lines has the propositional content that the lines are different lengths.[3] The Müller-Lyer illusion is an *illusion* because this propositional content is false.[4]

One straightforward interpretation of "there seems to be robust passage," at least when it is followed by "but this is an illusion," makes it also a claim about the content of some experience or experiences. It is the claim that the content of

Figure 11.1 The Müller-Lyer illusion.

---

[3] Some philosophers deny that "the lines look to me to have different lengths" and "my visual experience has the content that the lines have different lengths" are equivalent. They say that experiences cannot be categorized as veridical or non-veridical, that experiences do not have "propositional contents" that are true or false. I think these philosophers are wrong. See Byrne's "Experience and Content" for discussion.

[4] I will set aside the controversy over whether there are such things as veridical illusions.

some of our experiences includes the proposition that there is robust passage.[5] On this interpretation, attempts to explain why time seems to pass are attempts to explain why our experience has as part of its content the proposition that there is robust passage.

The argument, again, will be that experience favors the moving spotlight theory over the block universe because the moving spotlight theory has a better explanation. I call this the argument from the content of experience.[6]

This argument is one Schlesinger might have meant. Again, he wrote that "there is hardly any experience that seems more persistently and immediately given to us than the relentless flow of time." This is garbled; it seems to identify the flow of time with a kind of experience. Maybe what Schlesinger was trying to say was "it is immediately given to us in experience that there is robust

---

[5] Not all philosophers who say that it is just an illusion that there is robust passage mean that our experience falsely represents that there is robust passage. Smart just means that the way our language deals with time is easily misinterpreted, and when misinterpreted can mislead us into thinking that there is robust passage (see "Time and Becoming," around p. 11).

I thought at one point that necessarily all illusions were perceptual illusions. But the use of "illusion" to apply to "non-perceptual," or "cognitive," illusions is common in, for example, the work of the psychologist Daniel Kahneman. (His explanations of the cognitive illusions he has identified may be found in his book *Thinking, Fast and Slow*.) As best I can tell, the way Kahneman uses the term, not just any false belief is a cognitive illusion. Instead, a false belief counts as an illusion only if it is caused in a certain sort of way. A false belief is an instance of the focusing illusion, for example, when it results from paying too much attention to some things and thereby overemphasizing their importance, as when people pay too much attention to how much better the weather is in California, and thereby come to believe that those who live in California are happier. But I do not see an interesting argument for robust passage coming from the thought that, if the block universe theory is true, then the belief that time passes is a cognitive illusion. So I will continue to focus on the idea that the passage of time, if it is an illusion, is a perceptual illusion. Doing this also, I think, connects better with the intentions of most philosophers who discuss whether the passage of time is an illusion. Falk, for example, is very clear that when he says that the passage of time is an illusion, he means that it is a perceptual illusion. He is clear that he means that some of our perceptual experiences have the (false) content that time passes ("Time Plus the Whoosh and Whiz," pp. 211, 215). I think that most philosophers mean this, even if they are not as explicit as Falk. (Other places where it is relatively explicit that "illusion" means perceptual illusion are Davies, *About Time*, p. 275, and van Inwagen, *Metaphysics*, p. 79.)

Smart's claim that we are led to the false belief that time passes by misinterpreting our ways of talking about time appears in "The River of Time." In the later paper "Time and Becoming" he re-visits this claim and discusses several other possible "sources of the illusion." But none of the explanations he discusses make the illusion a perceptual illusion. Smart does not think any of these explanations are convincing, so I will not discuss them.

[6] The argument from content does not have "The moving spotlight theory is true" as its conclusion. Instead its conclusion is the "meta" claim that some fact favors this theory over the block universe theory. But let's not worry about the connection between the argument's actual conclusion and the conclusion we want.

Remarks that suggest the argument from content may be found in many places. Some places where it is relatively explicit are: Hestevold, "Passage and the Presence of Experience" (p. 541); Taylor, *Metaphysics* (p. 81); Smith, "The Phenomenology of A-Time" (p. 357); and Paul, "Temporal Experience" (pp. 334–9). (There may be alternative interpretations of some of these texts.)

passage." If we read "it is immediately given in experience that X" as "experience has the propositional content that X" we get the argument from the content of experience.

To begin evaluating this argument let us look at the assumption it starts with, namely the assumption that our experiences, some of them at least, have the content that there is robust passage. Now in the moving spotlight theory robust passage consists in change in which time is present. So if some of our experiences have the content that time passes, then some of our experiences have the content that some time is, or some things that exist in time are, present. And if some experience has the content that something is present, then, it seems safe to assume, some visual experience does. So the assumption the argument starts with entails that some things are visually represented as being present. Some things look like they are present. Which things? Presumably all the things I see. This apple, that chair, the blackboard—each does not just look to have a certain shape and a certain color. Each one also looks like it is present.

Before moving on, it is worth pausing to ask, even granting that this is true, how on earth could the moving spotlight theory even hope to explain it? The moving spotlight theory says that a time is present, not that apples, chairs, and blackboards are present. The answer is that the moving spotlight theory should not say that the only thing that is present is a time. It should say instead that things located at the present time are also present. Take that to be the version of the moving spotlight theory under discussion.

But we need to back up. Is it even true that apples, chairs, and blackboards look like they are present? I am not sure. It would help if I had some conception of what something has to look like in order for it to look like it is present. A ripe tomato and a firetruck look very different in many respects, but there is also a salient respect in which they look similar. And there is a respect in which both look different from oranges and chocolate bars. It is at least in part by attending to these similarities and differences that I get a grip on what something has to look like in order for it to look red. But the analogous procedure for getting a grip on what something has to look like in order for it to look like it is present does not work. For either everything or nothing I see looks like it is present.[7]

I will have to press on without this kind of conceptual help. Now "the tomato looks like it is present" strikes me as false. Actually, it strikes me as crazy, as absurd. But I do not want to place much weight on this judgment. For it may

---

[7] I suppose someone could say that in normal conditions everything I see looks like it is present, but there are very rare and very strange circumstances in which I can see something that does not look like it is present. But I do not know of anyone who has said this or said what those circumstances are, so this possibility is no help in the present context.

strike me as false only because I use the word "looks." Compare: does the tomato look like it exists? I find this question baffling. Nevertheless, I do think that my visual experience has the content that the tomato exists. Were I to learn that there are no tomatoes nearby I would think I was suffering from an illusion. Maybe, similarly, my visual experience represents that the tomato is present even though it sounds wrong to say that the tomato *looks* like it is present.

So far I have reached no conclusion about whether things look present. Here is one way one might try to argue for this claim. My visual experience has the content that the tomato is red. But "is" here is the present tense form of the verb "to be." And having the content that something X is (present tense) red is sufficient for having the content that X is present.

This is no good. The sentence "That tomato is red," where "is" is the present tense form, also has the content that that tomato is red. But if the block universe theory is true it does not thereby have the content that the tomato is present (in the non-reductive sense of "present" relevant to robust passage). And what we want right now is a reason to think that things are represented as (non-reductively) present that even a defender of the block universe can recognize.

Although I doubt that everything I see is visually represented as present, I am not going to rest my case on these doubts.[8] The argument from content is in trouble even if experiences do have this content.

Before we wade into that trouble I want to say one more thing about this claim that everything I see is visually represented as present. I arrived at it from the assumption that some experience represents that something is present. But moving spotlight theorists do not need to agree that some visual experience has this content. Of course, the moving spotlight theorist is hardly better off saying that it is only auditory experiences that represent time passing. How odd it would be to say that things sound like they are present but nothing looks like it is present. But they could maintain that no experience that comes to us from the five senses represents that something is present. Only experiences that come from introspection have this content. So, the thought is, it does not look like time passes, or sound like time passes. Apples do not look present and trains do not sound present. Instead we have experiences as of the passage of time only when we look "inward." It is our own internal states, the states that are the objects of introspection, that appear present. Some moving spotlight theorists might think that this is a more plausible claim than the claim that things look present.

---

[8] Some opponents of objective becoming are more willing than I to simply say that nothing looks present, deny that there is anything to explain, and rest their case. See for example Hestevold, "Passage and the Presence of Experience" (p. 542); Le Poidevin, *The Images of Time* (p. 78); and Callender, "Time's Ontic Voltage."

Introspection is a difficult topic. Some philosophers deny that we have a quasi-perceptual faculty of introspection, an "inner eye" with which we may gaze on our own internal states.[9] If our inner states do not appear to us any which way at all then they do not appear present and the move to introspection has not helped. But suppose this is wrong and we do have introspectable experiences. Then I am also willing to grant for the sake of argument that when I introspect an experience it is represented as present. However, in what follows I will stick to the case of visual experience. Nothing turns on making this choice.

So let us assume that experience does represent the occurrence of robust passage. Which theory of time better explains why this is so? It can look like the block universe is at a disadvantage. For it is surely easy to explain why there seems to be robust passage if there really is robust passage. It is as easy as using the fact that a tomato is red to explain why it looks red. So, the thought goes, the moving spotlight theory already has a relatively good explanation. We defenders of the block universe need an explanation that is at least as good. But what we have to explain is why our experiences represent things having a property that nothing can have. That has got to be hard to do. What, if the block universe theory is true, is the source of the illusion?

I only know of one attempt to answer this question. L. A. Paul claims that it is a necessary truth that experiences represent the things they are experiences of as (non-reductively) present.[10] I, however, doubt that this is so.

Still, as I have said, the argument from the content of experience is not good. We do not need to worry about how good an explanation the block universe theory has, because the moving spotlight theory's explanation is terrible.

Moving spotlight theorists want to explain why things look present by appealing to the fact that those things are present. So they think that presentness is a visible property, a property we can see something to have by looking at it. They think that presentness is, in this respect, like redness and squareness but unlike, say, electrical charge. Square things look square in part because they are square, but nothing looks negatively charged.[11]

---

[9] Byrne, in "Knowing What I See," building on some remarks by Evans in *The Varieties of Reference* (p. 226), defends this denial.

[10] She actually writes, "The idea is that the what-it's-like of an experience contains within it the experience as of nowness along with further experience (for example, as of redness). What it is to have an experience as of nowness is part of what it is to have an experience *simpliciter*" ("Temporal Experience," pp. 342–3). My reading of this passage is correct if an "experience as of nowness" represents things as being non-reductively present.

[11] On the "epistemic use" of "looks" something can look negatively charged. That is not the relevant use of "looks" here. Sometimes the relevant use is called the "phenomenal use" of "looks." See Byrne, "Experience and Content," for references on and discussion of this distinction. (Byrne denies

For their explanation of the fact that time seems to pass to be any good, moving spotlight theorists need to offer us some story explaining how, if their view is correct, presentness manages to be visible rather than invisible. Otherwise their explanation of the fact that time seems to pass has a big hole in the middle. I have never seen a moving spotlight theorist tell such a story.[12]

I have doubts about whether such a story can be told.[13] My doubts emerge from thinking about an argument for a different conclusion. D. H. Mellor argued that "we do not observe the tense of events," and some have read Mellor's argument as an argument that nothing looks present.[14] The argument does not establish this conclusion; that is why I did not give it above. But it leads us to where we now want to go. Consider this scenario. Jones looks through a telescope and sees a star. Unbeknownst to Jones the star is no more. It died in a supernova years ago. But the star was so far away when it died that the light from the supernova has not yet reached the Earth. Meanwhile Smith goes to the planetarium for a lecture on that star. The planetarium is so good that when Smith looks at the ceiling things looks exactly as they do to Jones as he looks through the telescope. Hestevold argues: "since past events [or past things] appear through the telescope to an observer [in this case, Jones] in the same way that present events [or things] appear to the observer [Smith]," nothing looks present ("Passage and the Presence of Experience," p. 541).

The problem with this argument is obvious. From the fact that the way things look is the same for Jones and for Smith, and the fact that nothing Jones sees is present, it does not follow that nothing looks present to either of them.[15] More carefully, it does not follow that neither person's visual experience has as part of its content the proposition that something is present. Maybe instead each of their visual experiences has this content.

If this is not obvious enough on its face consider the analogy with redness. Suppose Jones turn his telescope onto another star, one that still exists. That star

---

that existence of a phenomenal use of "looks," but still accepts the existence of a relevant use distinct from the epistemic use.)

[12] True, the theory has had few defenders. But that only explains why few have tried to explain how presentness manages to be visible. For reasons I am about to give, I think that even if many had tried they would have failed.

[13] Prosser, in "Could We Experience the Passage of Time?" (p. 88), gives an argument for this claim that in some respects resembles the one to follow. See also his paper "Passage and Perception."

[14] Mellor's argument is on page 26 of *Real Time*. I have modified the example and am presenting the argument as interpreted by others, not as Mellor understands it.

[15] There is another way to resist this argument. Externalists about content might deny that the way things look is the same for Jones and Smith, if "the way things look for Jones" refers to the propositions that are the contents of his visual experiences. (They will still admit that the scenes before their eyes are visually indistinguishable.)

looks red. In fact its color is indistinguishable from that of the tomato on his desk.[16] Unbeknownst to Jones the star is no longer red. Its color has changed as it has aged. But the star is so far away that the light indicating its current color has not yet reached the Earth. It would be wrong to argue that, since non-red things appear through the telescope to an observer in the same way as red things, nothing looks red.

The second star looks red to Jones, even though it is not red, in part because it was red when the light reaching him departed the star. The moving spotlight theorist might say something similar about the first star. He might say that the first star looks present, even though it is not, in part because it was present when the light reaching Jones departed the star.

But the analogy between redness and presentness falls apart when you think about it more. The picture in Figure 11.2 captures what is going on when the star looks red even though it no longer is. (In the diagram, gray represents red, black represents brown.) The star is 500 light years away. In 1511 the star is red and red light leaves the star. That light hits Jones' eyes in 2011. The star looks red to him.

What picture captures what is going on when the star looks present even though it no longer is? Maybe the picture in Figure 11.3? In that picture the halo indicates which time is, and which things are, present.

We are pursing the idea that presentness is visible for the same reason that redness is. Let's try substituting "present" in for "red" in the explanation we have just seen of the fact that redness is visible, while keeping an eye on Figures 11.2 and 11.3. The explanation for redness, again is this:

The star is 500 light years away. In 1511 the star is red and red light leaves the star. That light hits Jones' eyes in 2011. The star looks red to him.

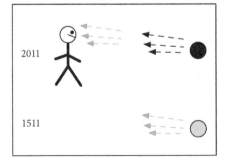

Figure 11.2 The brown star looks red.

---

[16] I do not spend much time looking at the stars. Perhaps no star could look the same in color as any tomato. If you think so then substitute a different example.

Figure 11.3 The star looks present?

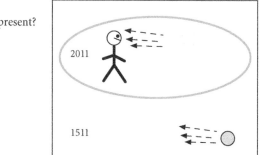

After making the substitution we get

The star is 500 light years away. In 1511 the star is present and present light leaves the star. That light hits Jones' eyes in 2011. The star looks present to him.

This is not correct. In Figure 11.3 no time in 1511 is present. Nor is anything in 1511, including the star, present.

We are running up against the feature of the moving spotlight theory that makes the theory so hard to grasp, namely the two-dimensionality of time. I need to be more careful in my use of tense and temporal language.

Let's work with MST-Supertense.[17] In that theory things and times super-are, or super-are not, present *simpliciter*. It is false that they super-are, or not, present at some times and not at others. So while the light super-leaves the star in 1511, it is not the case that the star super-is present in 1511. Neither super-is it the case that the star super-is not present in 1511. It super-is not present *simpliciter*.

Of course according to MST-Supertense the star super-was present. But it is false that "it super-was the case that P" super-is true if and only if there super-is some earlier time at which P. So we need to consider the picture in Figure 11.4 along with the picture in Figure 11.3. In the picture in Figure 11.4 the star super-is present. The picture in Figure 11.3 depicts what super-is the case; the picture in Figure 11.4 depicts what super-was the case.

But this only makes things worse. In one picture the star super-is present and in the other it super-is not. But the exact state of the photons leaving the star super-is the same in both pictures. Photons of the same frequencies depart in the same directions. And the exact state of the photons arriving at Jones's retina super-is the same in both pictures. It is hard to see how presentness can be a visible property if this is true.

---

[17] In what follows I will usually make my use of supertensed verbs explicit by prefixing them with "super-." I will not do this everywhere, though, since it makes for (and has made for, in earlier chapters) difficult reading. It should be clear which unmarked verbs are supertensed.

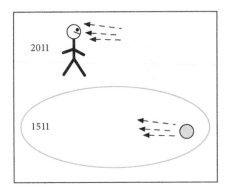

Figure 11.4 It super-was the case that . . .

Think about why redness is visible. Redness is visible because red things inter-
act with light differently from non-red things. (This is not to say that no non-red
thing ever looks red, just that this requires special circumstances.) This kind of
story about why presentness is visible has been ruled out. We have just seen that
being present makes no difference to the way something interacts with light.

Of course, in my argument that presentness makes no difference I assumed
that when the spotlight moves from one time to another nothing else changes.
Figures 11.3 and 11.4 differ only over the location of the halo. (More carefully,
I assumed that if P is a sentence that does not contain "present" or other A-
vocabulary, then if P super-is true it super-always is true, and conversely.) Might
a moving spotlight theorist abandon this assumption? He would have to say
something like, the haloed photons differ in some physical way from the un-
haloed photons. Maybe they have different frequencies. It is false that the photons
in 1511 super-are in the same physical state in both pictures. If the moving spotlight theorist says this then presentness does seem to be making a difference to
the light leaving objects that have it.

But it is hard to take this suggestion seriously. It seems flat-out inconsistent
with what we know about physics. In no physical theory that we might consult to
learn about the state of the light leaving that star does whether they are present
play a role in determining the photons' state.

So if presentness is visible it cannot be visible for the same kinds of reasons
that redness is visible.

This does not prove that presentness is not visible. For all I have shown it may
be visible for some other kind of reason. Suppose Jones has a twin brother. And
suppose that Smith does not know this and has never seen him. But Smith knows
Jones and sees him all the time. One day Jones's brother walks by Smith's win-
dow and Smith sees him for the first time. Smith certainly believes that Jones is
walking by. And Smith's visual experience certainly has the content that someone

with qualitative features F is walking by (where F are the qualitative features Jones has). But maybe more is true. Maybe Smith's extended acquaintance with Jones also makes it the case that his visual experience has the content *that Jones is walking by.* Perhaps the property of being Jones is represented in Smith's experience. If so then this property is visible even though Jones and his twin reflect light in just the same ways.

Maybe a story like this could be told about how presentness gets to be visible. I do not know how it would go. I would certainly like to see it.

Siegel, in *The Contents of Visual Experience,* argues that visual experience can represent properties much more complicated than shapes and colors. She argues that it can represent, among other things, causal relations and "kind" properties like the property of being a pine tree. Her preferred form of argument is the method of phenomenal contrast. A novice often walks in the forest and often sees pine trees. Later she studies dendrology and becomes an expert at identifying all kinds of trees including pines. The visual experiences she has when she looks at pine trees as an expert differ from the ones she had when she looked at them as a novice. The best explanation of this is that her expert experiences represent the trees as pines while her novice ones did not. But I do not see how to use the method of phenomenal contrast to argue that visual experiences represent things as present. What are the two contrasting experiences supposed to be?

Without a story about how presentness gets to be visible, the moving spotlight theorist does not have a better explanation of the fact that time seems to pass than does the believer in the block universe. The argument from content does not succeed.

11.3 The Argument from the Content of Experience, II

The argument from content is a two-edged sword. To the extent that it seems right to say that our experience represents that there is robust passage it also seems right to say that it represents that time is robustly passing at a certain rate. But, as we all know, time does not seem to pass at a constant rate. It seems to pass quite slowly while waiting in line at the DMV and much more quickly when something we enjoy is going on. These phenomena may not seem closely enough tied to the perception of the rate of time's passage, rather than after-the-fact estimates of how much time has passed. But other phenomena are more closely tied, for example the "oddball" effect. If a psychologist flashes red circles onto a screen, one after another, and then flashes something else, say a green circle, the oddball stimulus will seem to you to last longer, even if all the stimuli are flashed on the screen for the same amount of time.[18] But if the moving spotlight theory is true and we veridically perceive that things are present, and that there is change in which things are present, why do we misperceive the rate at which this change is occurring? The block universe theory does not appear to have much trouble explaining why time seems sometimes to speed up and slow down. (For example, Ian Phillips suggests, in his paper "Perceiving the Passage of Time," that when time seems to slow down it is because the subject is experiencing a greater-than-usual rate of "non-perceptual mental activity." This explanation is compatible with the truth of the block universe theory of time.)

Friends of objective becoming who think that the content of our experiences is the right place to look for an argument for their view might object that I have been looking at the wrong content. Maybe there is something else—not robust passage—that our experience represents, and its representing that is what favors the moving spotlight theory.

The most obvious candidate for this something else is change. One way to let time pass is to sit and watch the wheels go round and round. And if instructed to attend to the passage of time we normally just attend to the changes we are perceiving. We watch the second-hand move, or listen to the ticking of the clock.

Fine. While I doubt that our experience represents that anything is (non-reductively) present, it certainly represents things as changing.[19] This second argument from the content of experience is getting off to a better start. But where does it go from here? Why think that the fact that experiences sometimes represent change favors the moving spotlight theory over the block universe?

---

[18] See Pariyadath and Eagleman, "The Effect of Predictability on Subjective Duration." My favorite experiment on "subjective duration" is reported on in "Does time really slow down during a frightening event?", Stetson et al. (A description of it also appeared in a 2011 New Yorker profile on Eagleman.) Researchers strapped volunteers to a thrill ride that dropped them back-first one hundred and ten feet. (Into a net.) On their wrists was a watch with numbers zooming by too fast too read. Too fast to read in ordinary circumstances; the researchers wanted to see if as "time slowed down" during the frightening fall (subjects report the fall as lasting far longer than it actually does) the rate at which the numbers appeared to scroll by also slowed. If it did then they should be able to distinguish individual numbers as they fell. (They could not.) The correct philosophical interpretation of this research—what it suggests about the contents of our experience—is a difficult topic.

[19] Reid apparently challenged this claim. He held that our experiences always represent things as they are at an instant. Since change happens over time our experiences are silent about whether the things they represent are changing. What happens is that I remember that earlier the second-hand, or whatever, was in a different place than it looks to be now, and infer that it has moved. But this is implausible.

While it may be fairly uncontroversial that our experience represents change, how it does so is a matter of great controversy. Dainton surveys some options in "Time, Passage, and Immediate Experience" (this paper is also my source for Reid's view).

The only answer I can think of requires strengthening the premise:

When I say that our experience represents things as changing, I mean it represents them as *really* changing, as undergoing robust change. But there is no robust change in the block universe. So if the block universe theory is true then whenever something looks like it is changing this is an illusion. On the other hand, the moving spotlight theory can say that things look like they are changing (undergoing robust change) because they are.

I do not find this convincing at all. Why think that our experience represents things as undergoing robust change? I can see no reason. What's more, in the standard version of MST-Supertense the only kind of robust change is change in what is present. But our experiences do not ever represent this change. Even if things do look present they always look present. The changes our experiences do represent are not robust. So this version of the moving spotlight theory cannot just say that things seem to undergo robust change because they do.

There is a theory of time in which all change is robust change, namely presentism. But I am looking for arguments for robust passage, and the passage in presentism is not robust enough.

11.4 The Argument from the Phenomenal Character of Experience

Earlier I discussed Hestevold's argument that nothing looks present. But I cheated a little bit when I reported his conclusion. He did not write "nothing looks present." What he actually wrote is "*being present* is not a phenomenal property." Now "phenomenal property" has several uses in philosophy.[20] I am not sure which way Hestevold was using it. It may be that by "phenomenal property" he just meant a property that may be represented by experience. Then although he used different words his conclusion is the conclusion I reported: nothing looks present. But he might have meant something else by "phenomenal property." That something else suggests a different argument from experience.

Experiences do not just have representational content. They also have "phenomenal character." There is "something it is like" to have them. On one way of using "phenomenal property," talk about the phenomenal properties of an experience is talk about that experience's phenomenal character. When someone asks whether presentness is a phenomenal property, then, he may be asking whether what it is like to have an experience that is present differs from what

---

[20] Byrne explores the confusions around its use in "Sensory Qualities, Sensible Qualities, Sensational Qualities."

it is like to have one that is not. When Schlesinger said that the flow of time is "immediately given to us in experience" he may have been saying that we can tell that there is robust passage just by attending to our experience's phenomenal character.

Is there an argument in favor of the moving spotlight theory nearby? If there is it might go like this. The moving spotlight theorist starts by saying that, in his view, experiences that are present do differ in phenomenal character from experiences that are not. He then says that this allows his theory to explain things that the block universe theory cannot explain, or cannot explain well. This explanatory advantage, he continues, counts in favor of his theory. This, in outline, is the argument from the phenomenal character of experience.

Like the argument from the content of experience, the argument from phenomenology attempts to show that the moving spotlight theory is explanatorily superior to the block universe, and that for that reason we should believe it. The argument from phenomenology might seem good even in light of what I have said about the argument from content. Maybe the lesson of my discussion of the argument from content is that when arguing in favor of passage the best way to interpret "It seems like time passes" is not as a claim about the content of experience but instead as a claim about its phenomenal character.

I want to make another comment about the relationship between the argument from content and the argument from phenomenology. Many philosophers accept intentionalism, the thesis that the phenomenal character of an experience is determined by, or at least supervenes on, its representational content. If intentionalism is true then any moving spotlight theorist who says that experiences that are present have a distinctive phenomenal character must also say that they have a distinctive representational content. Does this mean that if intentionalism is true then the arguments from content and from phenomenology are the same argument?

Not necessarily. A moving spotlight theorist might say that although experiences that are present differ in content from experiences that are not, they do not differ over whether they represent anything as being present. Their content differs in some other way. (What other way? I have no idea. I am not suggesting that this is a plausible claim.) A moving spotlight theorist who says this rejects a premise of the argument from content. But he might still think that the argument from phenomenology was good. So that argument would still deserve independent consideration.[21]

---

[21] Many defenders of the block universe talk about the phenomenal character of experience when they talk about the fact that time seems to pass. This suggests the argument from phenomenology. Then in the next breath they say that although it seems like time passes this is just an illusion. This

Let us take a closer look at the argument from phenomenology. I have only presented it in outline. I have not said what it is that the moving spotlight theorist aims to explain. That is because I do not know any way of filling in this detail that makes the argument plausible.

What phenomenon might be explained? One candidate is the fact that my current experiences have *this* phenomenal character. Now this is not a particularly informative characterization of the phenomenal character of my current experiences. But it will do for now.

My current experiences occur on Tuesday. So my current experiences are my Tuesday experiences. Those experiences have a certain phenomenal character. The question is, why do those experiences have that character? The moving spotlight theorist says that they have it because they are present. The believer in the block universe will have some other explanation (it will not matter what it looks like).

Now as I phrased the question it looks like the moving spotlight theorist aims to explain every aspect of the phenomenal character of my Tuesday experiences. But he need not be that ambitious. It would do if he had a good explanation of just some aspect of their phenomenal character.

Okay, so, is the moving spotlight theory's explanation a better explanation of the phenomenal character of my Tuesday experiences? I am not going to answer this question directly. Instead I am going to discuss a related question. Is the moving spotlight theory's explanation any good? If it is not then we can safely assume it is not better than one a defender of the block universe can come up with.

If the moving spotlight theory's explanation is to be any good then the moving spotlight theorist has to say that experiences that are not present, but are otherwise as similar as possible to my Tuesday experiences, have a different phenomenal character. Is that right? To answer it will help to reflect on an example. So suppose that I spent today, Tuesday, in a red room (everything in

---

suggests the argument from content. Maybe they talk like this because they think the arguments are the same. Anyway, for this reason it is hard to give examples of philosophers who definitely have the argument from phenomenology rather than the argument from content in mind. One philosopher who does at least sometimes seem to aim just at the argument from phenomenology is Prosser in his paper "Could We Experience the Passage of Time?" He writes that the objective passage of time is not "directly perceived through the outer senses in the same way as colors." That looks like a rejection of a premise in the argument from content. Then he goes on to say that "time seeming to pass is a feature of conscious experience with a distinctive phenomenology" ("Could We Experience the Passage of Time?" pp. 76–7). That looks like an endorsement of the premise in the argument from phenomenology. (I think that in the first part of this quotation Prosser means to make the stronger claim that visual and auditory experience do not represent objective passage, not the weaker claim that either they do not represent this or they do but what they represent is false. I should say that this interpretation does not perfectly fit everything Prosser says. Some of the arguments in Prosser's paper, like the one I referred to in footnote 13, are aimed at the argument from content.)

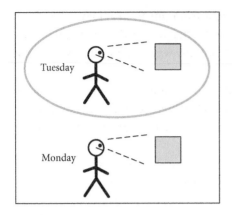

Figure 11.5 The red rooms according to MST-Supertense.

it is red); and I did so on Monday as well. On both days I have experiences as of red things, the same things on both days. If we continue to work with MST-Supertense, then the theory says that only the Tuesday experiences super-are present,[22] but the experiences super-are otherwise the same. Then the moving spotlight theory says that things super-are as in Figure 11.5.

Does the phenomenal character of the Monday experiences differ from that of the Tuesday experiences? I do not think so. I think that the Monday experiences have the same phenomenal character as the Tuesday experiences.

Suppose that I have a black-belt in autophenomenology. I can completely and accurately describe the phenomenal character of my experiences. (Perhaps I have invented a special language, more expressive than English, for doing this.) These descriptions are so complete that no one description correctly describes experiences that are phenomenally different. And my ability to appreciate the phenomenal character of my experiences is so good that I always produce an accurate description of the phenomenal character of the experience I am describing. Now suppose that on Monday I produce such a description, and I do so again on Tuesday. I write these descriptions down on paper. I think it is obvious that the descriptions will be word-for-word the same.

If I am right that the Monday and Tuesday experiences have the same phenomenal character then the fact that the Tuesday experiences are present cannot explain why they have the phenomenal character they have.

I know how a defender of the argument from phenomenology will reply to all this. She will deny that the Monday and Tuesday experiences super-have the same phenomenal character. What it super-is like to have the Tuesday experiences differs from what it super-is like to have the Monday experiences. To have

---

[22] I am speaking loosely. Only the experiences during one instant on Tuesday are present.

a way to name this difference let us say that on her view only the Tuesday experiences "super-feel lustrous." She will then say that the Tuesday experiences super-feel lustrous and the Monday experiences do not because only the Tuesday experiences super-are present.[23] And, the reply concludes, believers in the block universe cannot explain why only the Tuesday experiences feel lustrous.

There is more to this reply. Why am I inclined to think that the Monday and Tuesday experiences have the same phenomenal character? A moving spotlight theorist might say that I confuse this false claim with a true one. That is the claim that the phenomenal character the Tuesday experiences *super-have* is the same as the phenomenal character the Monday experiences *super-had*. That is, only the Tuesday experiences super-feel lustrous, but it super-was the case that only the Monday experiences super-feel lustrous. The phenomenal character of my experiences has changed. (On the block universe theory, of course, it cannot happen that my Monday experiences have a certain phenomenal character at one time (on Monday) and then a different phenomenal character at a later time (on Tuesday).)

What about the descriptions of the phenomenal character of my experiences that I wrote down? Since I have a black-belt in autophenomenology I know all about lustrousness. So I either wrote "lustrous" on both days or "lustrous" on neither day. ("Lustrous" is not the complete description; but it is the only part we are interested in now.) The moving spotlight theorist will probably say that I super-write "lustrous" on both days, even though only the Tuesday experiences super-feel lustrous. What then about the fact that I always produce accurate descriptions? She will say that understood one way this claim cannot possibly be true. For since the phenomenal character of a given experience changes, the accuracy of any description of it changes. No description of the phenomenal character of that experience can super-always be accurate. But there is another way to understand the claim so that it is true. Read it as the claim that I always produce descriptions that are accurate "when I produce them." That is, the description I super-write on Tuesday, the day that super-is present, super-is accurate, and the description I super-write on Monday super-was accurate. (But the description I super-write on Monday super-is inaccurate and the description I super-write on Tuesday super-was inaccurate.) That is as reliable as I can be about these things if MST-Supertense is true.

---

[23] By "E feels lustrous" a moving spotlight theorist might mean "E has some phenomenal character or other." Mellor seems to take this to be the view he is arguing against: "Once over a pain is no longer a pain, since nothing can be a pain, or an experience of any other kind, unless it is present" (*Real Time II*, p. 40). Peter Forrest defends a similar view in "The real but dead past," except that it is a version of the growing block theory, not the moving spotlight theory.

In spite of this reply I continue to think that my Monday and Tuesday experiences have the same phenomenal character. So I do not think that there are any facts about the phenomenal character of my experiences that the moving spotlight theory is better placed to explain than the block universe theory.

Why think that the Monday and Tuesday experiences are phenomenally the same? Consider the hypothesis that the difference between them is that my Tuesday experiences have some phenomenal character or other while the Monday ones have none. (In which case they shouldn't even be called experiences, but nevermind.) This theory says that throughout almost all of my life I am a (philosophical) zombie—a creature with no conscious experience. It says that in 1977 I super-have no conscious experiences, in 1994 I super-have no conscious experiences, and so on, except for one moment in 2013—at that time I do super-have conscious experiences.[24] I think this is wild. And I do not find it helpful to be told that I super-*had* conscious experiences in 1977. That does nothing to soften the blow. The theory still says that I super-am always a zombie throughout almost all of my life.

So I do not think that the argument from phenomenology is at all convincing. But I do think there is something interesting behind the argument. I just think that the argument from phenomenology is the wrong way to articulate that something. I think that when moving spotlight theorists are tempted to say that present experiences have a special phenomenal character it is because an entirely different argument from experience is pulling on them.

---

[24] Of course not even this will be true by the time that you read it, but something similar will.

12

Passage and Experience, II

12.1 The Argument from the Presented Experience

If their being present makes no difference to how things look, and my being present makes no difference to what it is like to see anything, then what possible connection could there be between experience and passage? Without either of these two claims, how is an argument from experience supposed to even get started?

There is a way. Go back to the beginning. Suppose, modifying my earlier example, that yesterday—Monday—I meditated with my eyes open in my white room and that today—Tuesday—I am meditating with my eyes open in my red room. The block universe theory says that I <u>see</u> white on Monday and I <u>see</u> red on Tuesday, and that is all (all that is relevant in this context anyway). So the picture of reality that goes with the block universe is the one in Figure 12.1 (gray represents red).

MST-Supertense says that this picture, regarded as a picture of how reality super-is, is incomplete. For one thing, it leaves out which time super-is present. So the picture should look like the picture in Figure 12.2. Again, the halo indicates which things super-are present. (And again, MST-Supertense says that the picture is still incomplete. Monday super-was present, but this fact is not evident in the picture. But that will not matter for now.)

Something about our experience is supposed to favor the second picture over the first. What about our experience? And how does that feature of our experience favor the second picture?

The third argument I want to discuss contains answers to these questions. Speaking for myself, I "look out on the world" from the "perspective" of just one time. But it is hard to see how this fact can be reconciled with the block universe theory. For that theory says that I exist at many times, times that make up a decades-long stretch of time. So what could make it true that I "look out on the world" from the perspective of only one of those times? It doesn't look like the block universe theory can answer this question. The moving spotlight theory, on

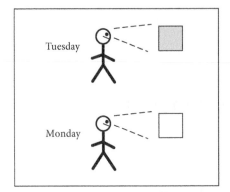

Figure 12.1 My meditations according to the block universe.

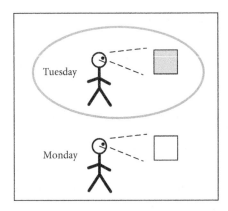

Figure 12.2 My meditations according to MST-Supertense.

the other hand, seems to provide an answer. That theory can say that I look out on the world from just this time because this time is the one that is objectively present.

That is a quick and dirty presentation of the argument. We need a statement of the argument that doesn't lean so hard on talk of "looking out on the world" from the "perspective" of a time. Here is a way to run the argument without them. It comes in two parts.

Stage 1. According to the block universe theory I <u>see</u> red on Tuesday and I <u>see</u> white on Monday. Let us use "the red experiences" for the visual experiences I have in the red room on Tuesday, and let us use "the white experiences" similarly. The block universe theory's characterization of the scenario seems to leave something out. Although I <u>see</u> white on Monday it is only the red experiences that are—and this is where things start to get tricky—*available to me* or *presented to me*.

Supposing this is true, why is it true? It is hard to see how to answer this question if the block universe theory is true. According to that theory

I am related to the red experiences and the white experiences in the same way. I <u>have</u> the red experiences at some time or other; the same goes for the white ones. So the theory cannot explain why only the red experiences are available to me by pointing to some difference between the way I am related to them and the way I am related to the white experiences.

The theory could in principle explain why only the red experiences are available by pointing to a difference in the experiences themselves. But what could the relevant difference be? True, the red experiences are experiences of red things and the white ones are not. But that cannot be why it is only the red ones that are available to me.

The theory's failure to explain why only the red experiences are available suggests that, if the theory is true, it is false that only the red experiences are available to me. Which is absurd.

The moving spotlight theory, on the other hand, can explain why only the red experiences are available to me. The spotlight of intrinsic privilege shines on the red experiences but not the white ones. The theory may draw a connection between being present and being available. It may say that the experiences of mine that are available to me, or presented to me, are all and only the ones that are present.

I have labeled this part of the argument stage 1 because, even if it is right, it does not establish the moving spotlight theory. Stage 1 aims to establish that exactly one time is present. But we also need some reason to believe that which time is present changes. The argument has a second stage that purports to do this.

Stage 2. Which experiences are available to me has been changing. Not only are the red experiences available to me, but, as I remember, the white experiences were available to me. Now suppose that stage one of the argument has been successful. The experiences available to me are the ones that are present. Then it follows that while Tuesday and things that happen on Tuesday are present, Monday and things that happened on Monday were present. But then which time is present has changed. Time has passed.

This is the argument from the presented experience.[1] It is certainly distinct from the other two arguments I have discussed. The argument does not say that things look present, or that the passage of time is represented by some experience. Nor does it say that being present makes a difference to the phenomenal character of any experience.

---

[1] As I said, I think that many arguments for the moving spotlight theory that appeal to experience are attempts to articulate this argument. Few do it very clearly. Ferré is onto something like it when

Furthermore, the argument is certainly an argument from experience. An analogous argument about a rock, rather than a subject of experience (namely me), has zero plausibility. Just try substituting "that rock" for every term referring to me in the first version of the argument:

> That rock "looks out on the world" from the "perspective" of just one time. But it is hard to see how this fact can be reconciled with the block universe theory. For that theory says that the rock exists at many times, times that make up a decades-long stretch of time. So what could make it true that the rock "looks out on the world" from the perspective of only one of those times? It doesn't look like the block universe theory can answer this question. The moving spotlight theory, on the other hand, seems to provide an answer. That theory can say that the rock looks out on the world from just this time because this time is the one that is objectively present.

The starting point of this argument, that the rock looks out on the world from the perspective of just one time, has nothing going for it. Rocks just do not have points of view. But "I look out on the world from the perspective of just one time" seems like just the right thing to say.

The argument from the presented experience is essentially first-personal. It is no datum for me that one of your experiences is distinguished, and vice versa. This is relevant to my discussion of relativistic versions of the moving spotlight theory in Chapter 9. It is not a flaw in a relativistic theory that it denies that both of our experiences are present, so long as it says that mine are.[2]

---

he names and discusses "the problem of temporal location" in "Grünbaum vs. Dobbs: The Need for Physical Transiency" (p. 280); see also his paper "Grünbaum on Temporal Becoming." Dainton hints at it when he says that "only sensations that are occurring now are actually felt" (*Time and Space*, p. 28). Balashov's "Times of Our Lives" is the most detailed and explicit presentation of something like the argument. (There are some differences: he sometimes focuses on the claim that I know that it is only my Tuesday experiences that are presented to me, rather than on the bare claim that they are presented to me.)

There is supposed to be a phenomenon that goes by the name "the presence of experience." The existence of this phenomenon is sometimes taken to favor robust passage. At the very least many defenders of the block universe feel called upon to discuss how this phenomenon may be reconciled with their theory. Now some descriptions of the presence of experience suggest the argument from the presented experience. But, as Balashov makes clear, when defenders of the block universe take up the task of reconciling the presence of experience with their view they rarely address this argument. Mellor in *Real Time II*, for example, ends up trying to explain why it is that whenever I have an experience I also believe that that experience is present (pp. 42–5). Whether this can be explained is also the focus of Hestevold's "Passage and the Presence of Experience" (see p. 543) and Oaklander's "On the Experience of Tenseless Time."

[2] The theory does need to say that when the present spacetime point falls inside my brain, then one of my experiences is present. I discussed the worry that the theory cannot do this in the first appendix to Chapter 9.

Let us get back to the argument. It will help to have a more explicit statement of the argument's premises. Stage 2 raises a lot of questions but I want to focus just on stage 1. The moving spotlight theorist and the block universe theorist agree that I have both white and red experiences (at different times). The moving spotlight theorist insists

(P1) Only the red experiences are available to me.

But, the moving spotlight theorist says, if the block universe theory is true then there is nothing in virtue of which it could be true that only the red, or only the white, experiences are available to me. So

(P2) If the block universe theory is true then either both the red and white experiences are available to me, or neither the red nor the white experiences are available to me.

From this it follows

(C) The block universe theory is false.

The same kind of argument does not work against the moving spotlight theory because the theory can be developed in a way that makes the analogue of (P2) false. The moving spotlight theory can say that all and only the experiences that are present are available to me. Since only the red experiences are present, only they are available.[3]

What is wrong with this argument? The argument makes use of a new and unexplained notion, namely the notion of an experience being available to someone. What is it for an experience to be available to someone? To answer this question we need to pay attention to tense. I have been playing fast and loose with it so far. Sentences like "Experience E is available to person P" play a key role in the argument; but how are we to interpret "is"?

We could read it as the present-tense form of "to be." Then the tenseless truth-condition for "E is available to P," the truth-condition the block universe theory gives, is:

- "Experience E is available to person P" <u>is</u> true at T if and only if person P <u>has</u> E at T.

To a certain extent this piles mysteries upon mysteries. For now we will want to know what it takes for it to be true that P <u>has</u> E at T. But if we focus on particular

---

[3] A similar argument may seem to support presentism over the block universe theory. My reply to the argument, when formulated as an argument for presentism, would be the same as the one I am about to give.

examples this notion need not be mysterious. If P <u>sees</u> a tree at T, or <u>hallucinates</u> a tree at T, then P <u>has</u> a visual-treeish-experience at T.

We could also read the "is" in "E is available to P" as the tenseless form of "to be." Now fully explicit tenseless talk of availability must be time relative. So the tenseless truth-conditions for tenseless availability-talk is:

- "Experience E <u>is</u> available to P at T" <u>is</u> true if and only if P <u>has</u> E at T.

(One may of course leave out a time reference, and merely say something of the form "E <u>is</u> available to P," as long as it is understood from the context what the relevant time is.)

Now the argument from the presented experience will not look interestingly different if we read the talk of availability as tensed or tenseless. Since the tenseless reading makes things clearer let's look at it. (P1) is uttered on Tuesday, so it becomes

(P1.1) Only the red experiences <u>are</u> available to me on Tuesday.

But now look at the analogous reading of (P2):

(P2.1) If the block universe theory is true then either both the red and white experiences <u>are</u> available to me on Tuesday, or neither the red nor the white experiences <u>are</u> available to me on Tuesday.

That's not right. It is false that the white experiences <u>are</u> available to me on Tuesday. The argument for (P2) was that there are no relevant differences between the experiences or my relations to them. But the argument was looking in the wrong place. The difference in availability is grounded in a difference in the relation each experience bears to the time at which (or the context in which) I ask, or wonder, which experiences are available to me.

Moving spotlight theorists will insist that this is not how they intended the argument to be understood. The talk of availability the argument employs, they will say, exhibits neither ordinary tense nor time-relativity.[4]

It will not do for the moving spotlight theorist to insist on more than this. They should not insist that "is," in "E is available to P," is in the present supertense.

---

[4] You can see the struggle to distinguish a version of the argument that uses a time-relative notion of availability from a version that uses a time-independent notion in Craig's response to Baker's response to Ferré (Craig, *The Tenseless Theory of Time*, p. 169; Baker, "Temporal Becoming: The Argument from Physics"; Ferré, "Grünbaum vs. Dobbs"). Balashov, in "Times of Our Lives," also works to isolate the second interpretation of the argument from the first.

Interpreted the second way, the argument from the presented experience resembles David Lewis's well-known argument from temporary intrinsics for the perdurance theory (*On the Plurality of Worlds*, pp. 202–4). I will say something more about this connection in footnote 20.

Then the argument can only hope to convince someone who already recognizes the existence of supertensed verbs. But even those who accept the block universe theory accept the existence of verb forms that do not exhibit ordinary tense. For they accept the existence of tenseless verb forms.

We need a way to mark verbs that do not exhibit ordinary tense but which still may be either tenseless or supertensed. I will do this by putting them in capitals. So the intended reading of (P1) is

(P1.2) Only the red experiences ARE available to me simpliciter.

The "simpliciter" indicates a lack of time-relativity. Similarly, the intended reading of the second premise is

(P2.2) If the block universe theory is true then either both the red and white experiences ARE available to me simpliciter, or neither the red nor the white experiences ARE available to me simpliciter.

In response to this a defender of the block universe might put her foot down and say that it makes no sense to talk about which experiences are available to someone without using ordinary tensed verbs and without time-relativization. The argument fails because both premises (P1.2) and (P2.2) falsely presuppose that this does make sense. "Only the red experiences ARE available to me simpliciter" may seem like a tempting thing to say, the idea goes, but only if it is confused with the tensed sentence "Only the red experiences are available to me." (This second sentence is true on Tuesday, the day I sit around contemplating which experiences are available to me.)

12.2 Spotlight vs. Spotlight

We who believe the block universe theory are not forced to say that "experience E IS available to person P simpliciter" is nonsense. We can always accept its intelligibility and deny premise (P1.2). All the experiences I ever have ARE available to me, not just the ones I am presently having. This, however, can sound quite counterintuitive. So it can look like the best thing to do is to say that this kind of availability-talk is nonsense.

But there is another option.

To build suspense I am going to hold off discussing it until the next section. In this chapter the moving spotlight theory has morphed into something new. It is worth pausing to wrap our minds around the version of the theory we are now working with.

The version of MST-Supertense that played a starring role in Chapters 6 and 7 says that there is robust change in which time super-is present, and in nothing else. But the idea that the best version of the theory will have other kinds of robust change in it than this has come up repeatedly. It came up most recently in the discussion of the argument from phenomenology. There the suggestion was that individual experiences undergo robust change. There are experiences that super-have no phenomenal character but super-will have some phenomenal character (namely, those that are located at times later than the present time).

The version currently on the table also says that things other than time undergo robust change. But it does not say that the other things are experiences. It says that the other things that undergo robust change are people (and other conscious things), like you and me. We undergo robust change in which of our experiences super-are available.

This can be put without using the words "experience" and "available." The theory does not just say

(1) BAS super-sees a white wall on Monday.
(2) BAS super-sees a red wall on Tuesday.

and so on; it also says

(3) BAS super-sees a red wall.

That is, it says that BAS super-sees a red wall "full-stop." In the theory "BAS super-sees a red wall" is metaphysically complete. There is of course a connection between (2) and (3): (3) super-is true because (2) super-is true and Tuesday super-is present.[5]

Now I dismissed the argument from phenomenology without a lot of hand-wringing. But I am taking the argument from the presented experience quite seriously. Why? Is there really that much difference between the versions of the moving spotlight theory the arguments aim to establish?

I think there is. The version at work in the argument from phenomenology says that I am almost always a zombie. Although I exist at many times I only have conscious experiences at the time that super-is present. I recoil from this theory in horror. Try to imagine the spatial analogue. Right now, all people but one (me I hope) are zombies. I just do not think that is the world I live in.

One could try to improve the theory by removing the zombies. Have it say, not that I am zombie at all times but the present, but that I do not exist at all at times other than the present. Then the theory turns into the theory of objective becoming in which we move through time, which I described in Chapter 10. I expressed some reservations about that theory there.

---

[5] Sider discusses a theory like this on pp. 260–1 of *Writing the Book of the World*.

So let's focus on the new version of the moving spotlight theory. In it, although I exist at many times, my mental life does not really unfold over time. It is closer to the truth to say that it unfolds in supertime, except that there is no supertime.

The theory takes the block universe and adds a fact about which time super-is present. This fact then determines which experiences super-are available to me. The picture we get is the one in Figure 12.3, in which "presentness" passes through spacetime from past to future and the box indicating which experiences super-are available to me marches up my worldline.

If the moving spotlight theory is true, a common objection goes, then we cannot know what time is present.[6] Suppose I look at my watch every hour on the hour and each time think that the time the numeral on my watch represents is present. Then, it seems, even if my watch is perfectly accurate I am almost always wrong. For there I am, thinking, at T, that time T is present, for numerous values of T. But at most one of those thoughts is correct.

This argument does not touch the current version of the moving spotlight theory. That version says that the thought that X o'clock is present is only presented to me, is only a thought that I super-am having, when X super-is present. So whenever I have a thought like this it is true.[7]

This new version of the moving spotlight theory resembles the theory from Chapter 10 in some respects. I do not move up through spacetime, true, but my

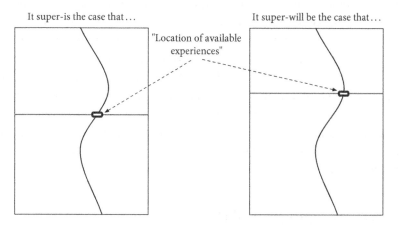

It super-is the case that...   It super-will be the case that...

"Location of available experiences"

Figure 12.3 MST-Supertense with time-independent availability.

---

[6] Bourne takes this argument to refute every theory of robust passage except presentism; see *A Future for Presentism*.

[7] The "whenever" in this sentence does not mean "at any time." Instead, "Whenever X, Y" as used here means "It super-is always the case that if X, Y."

Sider presents a similar response to this argument on behalf of a somewhat different theory of robust passage on page 261 of *Writing the Book of the World*.

"awareness" does. In fact this theory seems to capture what the theory in Chapter 10 was supposed to capture, but better. That theory was supposed to capture the idea that as we sail on the ocean of time we come across events. But in the theory there are no events out there in the future waiting for us to come across them. The standard moving spotlight theory, on the other hand, does have events out there in the future; but there is no sense in which we are moving toward them. In the current version there is.

The theory I have been discussing is a version of MST-Supertense. Is there an analogous version of MST-Time? The original version of MST-Time said

- From the perspective of Monday, I see white on Monday and red on Tuesday.
- From the perspective of Tuesday, I (also) see white on Monday and red on Tuesday.

The new version should say

- From the perspective of Monday, I see white.
- From the perspective of Tuesday, I see red.

(It will also say that from each perspective I see white on Monday and red on Tuesday.) But when we take the argument from the presented experience seriously, this theory is no good. That argument aims to establish that my seeing red is distinguished in some way; but in this version of MST-Time, just as in the block universe theory, it has too much in common with my seeing white.

In the end, then, I agree with the criticism that theories like MST-Time are bad because in them no single time is privileged. But I do not think that theories like this are bad just because it is some kind of a priori truth that robust passage requires a privileged time. I think they are bad because they are not supported by the best argument from experience. (These reasons to reject MST-Time do not apply to the version of MST-Time with absolute presentness from the first appendix to Chapter 4.)

12.3 Accepting Availability

Let's get back to the main road. I said that declaring the argument from the presented experience to be nonsense is not our only hope. There is another. There is a presupposition of the discussion in Section 12.1 that a believer in the block universe theory could reject. I have in mind the claim that the block universe theory "says that I exist at many times, times that make up a decades-long stretch of time." The block universe theory alone does not say this. It says this only in

Figure 12.4 The stage theory.

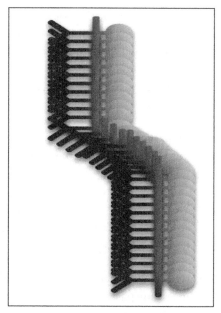

conjunction with the endurance, or the perdurance, theory of persistence. But there is another, quite radical, theory of persistence that denies that I exist at a decades-long stretch of time: the stage theory of persistence.

The stage theory says that I exist at just one time.[8] We can re-interpret the left panel of Figure 10.2 to illustrate this part of the stage theory. That panel is reproduced here as Figure 12.4. Let each human-shaped thing in the figure represent an instantaneous person-stage. The perdurance theory says that I am the four-dimensional thing that has those person-stages as parts. The stage theory says instead that I am identical to just one of those person-stages. What the perdurance theory says are earlier parts of me the stage theory says are other people who closely resemble me. (And the theory does not just say this about me. It says that all people, all tables, all chairs, indeed everything that we talk about or refer to in ordinary, non-philosophical contexts, is a thing that exists at just one time.)

I will get to how the stage theory helps with the argument from the presented experience in a minute. First we should ask whether the cure is worse than the disease. How can a theory that says that I exist at only one time (and is set in the block universe theory) be anything but obviously false?

The stage view is not obviously false. The theory does not say, for example, that obviously true sentences like "Jane was once four feet tall" are false. Why not? Well, the falsity of "Jane was once four feet tall" follows from the claim that

---

[8] The stage theory was first proposed and defended by Sider in "All The World's A Stage."

Jane exists at only one time only if a certain analysis of *de re*-tensed predication is correct. There is only trouble if we assume that the analysis of simple past-tensed predications looks like this:

- "X was F" is true at T if and only if X exists at some time S earlier than T and X is F at S.

But it is part of the stage view that this analysis is false. The stage theory endorses, instead, temporal counterpart theory. The counterpart-theoretic analysis of *de re*-tensed predication, applied to our example sentence-form, looks like this:

- "X was F" is true at T if and only if X has a temporal counterpart (that of course exists for just an instant) that (i) is temporally located at a time earlier than T and (ii) is F.

Clearly, on this analysis, Jane can satisfy "X was once four feet tall" without existing at any past times.

Who are my temporal counterparts? There may be no definitive answer; one version of the stage theory says that in different contexts "X and Y are temporal counterparts" expresses different relations. Different views about the "persistence conditions" for human persons correspond to different candidate counterpart relations. One candidate relation, for example, has it that my counterparts are the people who are physically and psychologically continuous with me in the right way. (The person-stages that, if the perdurance theory is true, are my instantaneous temporal parts are, in the stage theory, my temporal *counter*parts.) This package of views—the claim that no person exists at more than one time and temporal counterpart theory—is the stage theory.

I have said that the stage theory does not have the bad feature that obviously true-tensed sentences like "Jane was once four feet tall," or "I have existed for many years," come out false in the theory; they come out true. Some continue to think that the theory is obviously false. It is, after all, part of the theory that I exist at just one instant. Isn't this obviously false? Isn't it obvious that I exist at many past instants?[9]

Let's not talk about me. I doubt it is obvious to those with no exposure to the philosophy of time that they exist at many past instants, since I doubt that they have ever encountered tenseless language.[10] So I doubt that they could make much sense of the sentence "You exist at many past instants."

---

[9] This is Merricks's objection to the stage theory, from his review of Hawley's *How Things Persist*.

[10] Maybe I should also exclude those who have studied spacetime physics.

What about the philosophers? Probably plenty of philosophers think that they <u>exist</u> at many past instants. But I doubt that this belief of theirs is a basic belief.[11] I suspect that it is instead based on the combination of: their (tensed) belief that they have existed for a long time, and their acceptance of a particular version of reductionism about tense. The stage theory is an alternative version of reductionism about tense. Is it really less obvious than the "standard" version? It seems to me that what is obvious is the tensed claim that we have all existed for a while. When it comes to which of these two reductive analyses of tense is correct, nothing is obvious.[12]

This is only a partial defense of the stage theory. A comprehensive defense of it is not my aim.[13] My aim is just to explain how embracing the stage theory allows me to respond to the argument from the presented experience.

So let me now turn to that topic.[14] Look again at Figure 12.1, reproduced here as Figure 12.5. The pointed question directed at the block universe was, why does it seem that only my Tuesday experiences ARE available to me? If "E IS available to P" is not a time-relative notion then there does not seem to be any difference between my Monday and Tuesday experiences, or in my relation to them, to use in an answer.

This question just goes away if the stage theory is true. The stage theory says I <u>exist</u> on Tuesday but not on Monday. So if the stage theory is true then Figure 12.5

[11] And if it is basic, I think they are being irrational.

[12] Sometimes I am willing to reject an analysis out of hand, even if "all the right sentences come out true" under that analysis. That is how I feel about modal realism. When is this okay, and when is the truth of the analysis too theoretical for one's pre-theoretical beliefs to have any bearing on it? I wish I knew.

I should also say that stage theorists need not say that it is false in every sense that any of us <u>exists</u> at more than one time. They could say that "X <u>exists</u> at T" is ambiguous. In one sense, no person <u>exists</u> at more than one time. But in another sense, "X <u>exists</u> at T" <u>is</u> true if and only if X <u>has</u> a temporal counterpart that <u>is</u> temporally located at T. In this second sense, each of us <u>exists</u> at many times. They could go on to say that it is the second sense that "exists at" has when people affirm that they exist at many times.

The distinction between the two senses of "<u>exist</u> at" parallels David Lewis's distinction between existing in a possible world and existing according to a possible world. Lewis draws this distinction in *On the Plurality of Worlds* (p. 96). He uses it to defend modal counterpart theory against a similar objection.

I am not sure there is much value in multiplying senses of "<u>exists</u> at." I think it is enough to say that when people say "I <u>exist</u> at many times" or "I exist at many possible worlds" they are presupposing certain theories of time or of modality, and that what they really believe is what you get by subtracting the presupposition from the content of the sentence they utter. In the case of "I <u>exist</u> at many times" what is left is the tensed belief. (Yablo presents a comprehensive theory of "logical subtraction" in his book *Aboutness*.)

[13] More comprehensive defenses may be found in: Sider's "All The World's A Stage" and *Four-Dimensionalism,* and Hawley's *How Things Persist.*

[14] The idea of appealing to the stage theory to defend the block universe against an argument like this appears in Balashov's "Times of Our Lives."

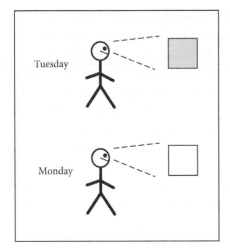

Figure 12.5 My meditations according to the block universe.

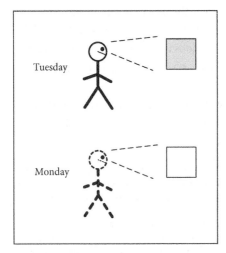

Figure 12.6 My meditations according to the block universe theory + the stage theory.

is wrong. We need to consult instead Figure 12.6. In that figure the stick figure drawn solid is me. The one drawn dashed is someone else.

What gives the question why only my Tuesday experiences ARE available to me its bite is the assumption that I have the red experiences on Tuesday and also have the white experiences on Monday. Given this it seems impossible that I could have, in a non-time-relative way, only the red experiences. But if the stage theory is true then I do not exist on Monday in the first place, and so do not have any experiences on Monday. If I do not have any white experiences on Monday, there is no mystery about why there are experiences that I have at some time that are not available to me in a non-time-relative way.

So if the stage theory is true then accepting that it makes sense to ask which experiences are available to me *simpliciter* does not land the block universe theory in hot water. The theory can say that all the experiences someone <u>has</u> are available to that person. Since I <u>exist</u> only on Tuesday those are the only experiences that are available to me. The argument from the presented experience is defeated.

12.4 Evaluating the Options

So, faced with the argument from the presented experience, we block universe theorists may reject "experience E IS available to person P" as unintelligible, or we may embrace the stage theory. Which is the better option?

Adopting the stage theory might seem like an extreme measure. I nevertheless used to think that it was the better option. I thought I had an argument that we should accept that "experience E IS available to person P" is intelligible. I no longer find that argument persuasive. But I still think that reflecting on the scenario that was the focus of that argument is a good idea, because it helps further undermine the argument from the presented experience.[15]

So consider the spatial analogue of the meditation scenario. In the original scenario I see different colors on different days. I see only red things on Tuesday and only white things on Monday. What might the spatial analogue of this scenario be? It will have to be one in which I simultaneously see different colors in different places.

So in the new story I will be simultaneously located in two different places. Of course, I am already simultaneously located in two different places. I am over here, where my left hand is, and I am also over there, where my right hand is. But anyone who looked in either of those places would see only a hand. In the scenario I have in mind something far more spectacular is going on. I am meditating in a red room in Massachusetts while simultaneously meditating in a white room in Florida. Anyone who walked into the red room would say he saw a whole person: head, torso, arms, legs, and so on. Anyone who walked into the white room would say the same.

(Do I manage to be both in Massachusetts and in Florida by having one human-shaped part wholly in Massachusetts and another human-shaped part wholly in Florida? Or do I have just one human-shaped part that is simultaneously wholly located both in Massachusetts and Florida? One set of answers (no and yes) gives the spatial analogue of the endurance theory; the other set is the spatial analogue of the perdurance theory. It does not matter which view is true in the story.)

---

[15] The original argument appeared in "Experience and the Passage of Time."

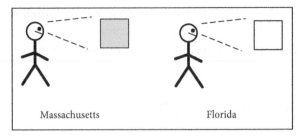

Figure 12.7 My meditations according to the spatial block universe.

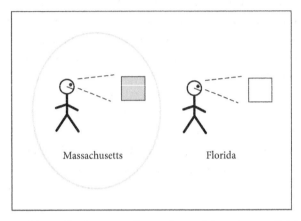

Figure 12.8 My meditations according to the spatial moving spotlight theory.

According to the "spatial block universe," we can say everything there is to say about this scenario by saying what experiences I am having in which places. The picture of reality that goes with this theory is the one in Figure 12.7. The spatial version of MST-Supertense agrees that I see red things, have red experiences, in Massachusetts and that I see white things, have white experiences, in Florida. But it goes on to say that one and only one place is present.[16] The halo in Figure 12.8 indicates that in this scenario it is my location in Massachusetts that is present.

The spatial analogue of MST-Supertense does not just endorse the picture in Figure 12.8. It also says that there is some spatial analogue of supertense. Although only a place in Massachusetts is present, "to the south" only some place in Florida is present. This is all very mysterious. Fortunately, I think we can ignore this part of the theory.

[16] Maybe the spatial moving spotlight theory should not use "present" for the special property that privileges just one location. Maybe "present" has too many temporal connotations. Nevertheless, I will stick with it.

Now, which of the spatial pictures is better? It is hard to say, because both are ridiculous. Both say that I am simultaneously in Massachusetts and in Florida. But I am not; I am only in Massachusetts.

Still, suppose that I have become convinced that I am simultaneously in Massachusetts and in Florida. Perhaps the oracle of philosophy revealed it to me. And suppose you have become convinced of something similar, and are contemplating a similar pair of pictures depicting you. So we accept the common presupposition of the spatial block universe and the spatial moving spotlight theory. Now which theory is better?

Again the debate will focus on a question about which experiences are available to me. The spatial moving spotlight theorist will say that we see red in Massachusetts and white in Florida. But there must be more to it than that. I think that it is only my red experiences that are available to me. Surely you think the same. The spatial spotlight theory captures these facts and the spatial block universe theory does not.

Now look at the response to this argument that parallels the first response to the argument from the presented experience. "Metaphysically complete" statements about which experiences are available to me must be of the form "E is available to me at location L." (One may still say things of the form "E is available to me," as long as it is understood from the context what the relevant location is.) So it makes no sense to talk about which experiences are "available to me simpliciter," where the "simpliciter" indicates a lack of location-relativity. Talk of an experience's being available is only intelligible if it is location-relative. All we can say is that the red experiences are available to me in Massachusetts and the white ones are available to me in Florida. Insofar as I am tempted to say that the red ones are available to me simpliciter, the response goes, it is because "The red experiences are available to me" is true in Massachusetts (if said by me), which is the place where this conversation is taking place.

I used to find this difficult to believe, but now it sounds just right, and the argument for the spatial moving spotlight theory seems devoid of force. The brain activities that "underwrite" my contemplation occur in Massachusetts. No wonder then that the experiences that I want to say are available are the ones that are also underwritten by brain activities in Massachusetts. To the extent that I succeed in imagining myself divided, living two lives in different places, it is no surprise that the experiences I have in one of those places are inaccessible to me in the other place.

Aren't we willing to say something similar about much less far-fetched scenarios? There are well-known experiments in which "split-brain" subjects, subjects who have had their corpus callosum severed, have a scene presented to them so that visual information about the left half of the scene is available only to the right hemisphere of their brain, and vice versa. Say, the word "key" is on the left and the word "ring" is on the right. Asked to identify the word, the subject will say that he sees only the word "ring," but will pick up a key with his left hand.[17] It is natural to say that the subject sees "ring" "on the left" and sees "key" "on the right."[18]

Imagining what our experience would be like were we spatially scattered—or, at least, if successful imagining seems unlikely, thinking through what we should say about our experience in a scenario in which we are spatially scattered—may help break down resistance[19] to saying the analogous thing about the temporal case. Assuming that the stage theory is false, the block universe theory says that we are temporally scattered. My having experiences at other times that are inaccessible, or unavailable, to me now is no stranger or more surprising that my having experiences in other places that are inaccessible to me here.

12.5 Appendix: Availability without the Stage Theory?

Must we accept the stage theory if we find "experience E IS available to P simpliciter" intelligible, and we like premise (P1.2) of the argument from the presented experience?

One who accepts the perdurance theory can accept that there are some things to which experiences <u>are</u> available simpliciter, namely instantaneous stages of persons. All and only the experiences a stage <u>has are</u> available to it. But since the theory identifies persons with four-dimensional spacetime worms rather than stages, no experiences <u>are</u> available simpliciter to me, or to persons generally. So if we accept this theory we will have to reject (P1.2). But maybe we can use the fact that, in the perdurance theory, "experience E IS available to P" is intelligible, to explain the temptation to accept this premise. (Of course, not everyone will think there is a temptation here to explain.)

How might the explanation go? Putting it in the first person, I could note that on Tuesday, my Tuesday stages "do my thinking for me." And the red experiences <u>are</u> available (simpliciter) to my Tuesday stages, so an analogue of (P1.2) is true of those stages. That is why (P1.2) is tempting.

I am not really sure what to think about this. Lots of things are true of my stages and not of me. My stages are instantaneous. There are times S and T such that my

---

[17] See Sperry, "Lateral specialization," or Bayne, *The Unity of Consciousness*, chapter 9, for more detail on these experiments.

[18] I should say that Bayne, for one, resists this common interpretation of the data. On his view, at any time, the subject either sees "key" or sees "ring," but not both. Which he sees is sensitive to the circumstances. Evaluating his arguments is beyond the scope of this book.

[19] If you had any.

S-stage has never been happy and my T-stage will never feel regret. But I do not confuse these properties of my stages with properties of me. At S I do not think I am instantaneous, or that I have never been happy, and at T I do not think I will never feel regret. My S-stage does not think the thought "I am instantaneous" or the thought "I have never been happy." Why should it be different with "These experiences <u>are</u> available simpliciter to me"?

Maybe the idea is that acquaintance with the red experiences reveals somehow that they <u>are</u> available simpliciter to something, and I too quickly conclude that it is me they <u>are</u> available to. I think this explanation is also problematic. It says that I can know that these experiences <u>are</u> available to something, without knowing that they <u>are</u> available to me. Is that really possible?[20]

I am not saying that the perdurance theory cannot explain the appeal of premise (P1.2). The theory can always say what the endurance theory can also say: that we confuse "The red experiences <u>are</u> available to me simpliciter" with "The red experiences are available to me on Tuesday," and we do this because we abbreviate the second thought as "The red experiences are (presently) available to me," and think that thought on Tuesday. What I doubt is that the perdurance theory can use the fact that, in his theory, talk of an experience's <u>being</u> available to something simpliciter is intelligible, to explain our temptation to accept (P1.2).

12.6 Appendix: Thank Goodness That's Over

The second-most famous argument in the philosophy of time, after McTaggart's, is probably Prior's "Thank Goodness That's Over" argument. This argument is sometimes classified as an argument from experience for robust passage, but in fact conscious experience plays no important role in it.[21] And there are already excellent responses to the argument by block universe theorists out there in the literature. So I did not say anything about the argument in the main text. Still, there is an interesting connection between one popular response to the argument and the stage theory, so I want to say something about the argument in this appendix.

[20] The dialectic here resembles that surrounding David Lewis's argument from temporary intrinsics for the doctrine of temporal parts. It seemed at first that Lewis's argument against endurantism was that endurantism falsely entails that no tennis ball (for example) is round simpliciter. But that is also false on perdurantism, his preferred theory. Lewis corrected this impression in "Rearrangements of particles: reply to Lowe." His premise is just that some things are round simpliciter. Perdurantism is compatible with this premise but endurantism is not. To defend this response one needs to explain how someone can come to know that some things are round simpliciter without knowing that tennis balls, or baseballs, and so on, are round simpliciter.

[21] D. H. Mellor discusses the argument in a section of *Real Time II* called "The Presence of Experience," and a well-known anthology in the philosophy of time, Oaklander and Smith's *The New Theory of Time*, places discussions of Prior's argument in a section headed "Time and Experience."

Prior first presented his argument in this passage:

One says, e.g., "Thank goodness that's over!", and not only is this, when said, quite clear without any date appended, but it says something which it is impossible that any use of a tenseless copula with a date should convey. It certainly doesn't mean the same as, e.g. "Thank goodness the date of the conclusion of that thing is Friday, June 15, 1954", even if it be said then. (Nor, for that matter, does it mean "Thank goodness the conclusion of that thing is contemporaneous with this utterance". Why should anyone thank goodness for that?) ("Thank Goodness That's Over," p. 17.)

Prior's target here is quite clearly the block universe theory—in particular, reductionism about tense. He claims that the sentence he made famous is not equivalent to any tenseless and A-vocabulary free sentence.

There are several ways to reconstruct Prior's argument. Here is a way that uses a minimum of technical notions. Change the example and suppose that Jane has just finished a marathon. Looking at her face we can tell that

(1) Jane is relieved that the race is in the past.

That is the set up. The argument itself has two premises:

(2) If the block universe theory is true then (1) "says something" which could also be conveyed by a tenseless and A-vocabulary-free sentence.
(3) But (1) "says something which it is impossible that" any tenseless and A-vocabulary-free sentence "should convey."
(4) Therefore, the block universe theory is false.

Prior's talk of what a sentence "says" suggests that we understand (2) to say that if the block universe theory is true then (1) expresses a proposition that cannot be expressed by any tenseless and A-vocabulary-free sentence.[22] But I am avoiding proposition-talk in this book so for now I will put the argument in terms of truth-conditions:

---

[22] Prior's words, again, were that (1) "says something" that no tenseless and A-vocabulary-free sentence could "convey." While talk of what a sentence "says" is naturally understood as talk of what proposition that sentence expresses, talk of what a use of the sentence "conveys" is naturally understood as talk of either (i) what proposition a speaker asserts, on a given occasion, by saying that sentence, or (ii) what proposition a hearer comes to believe, on a given occasion, by hearing someone say that sentence. On some views, speakers do not always assert the propositions expressed by the sentences they utter. Then (1) might have a reductive truth-condition, even if it can be used to assert something that no tenseless and A-vocabulary-free sentence can be used to assert. I am not drawn to this response, so I will ignore the distinction on which it is based. (I will address a related response below, namely the idea that the proposition, or "centered proposition," that is said by (1) to be the object of Jane's relief, is not the proposition expressed, in the context in which (1) is used, by "The race is in the past.")

(5) If the block universe theory is true then (1) has a tenseless and A-vocabulary-free truth-condition.

(6) But it does not.

(7) Therefore, the block universe theory is false.

Why believe (6)? Prior does not give an argument. He just asserts that candidate truth-conditions like (8) and (9) do not work:

(8) Jane is relieved that the end of the race is earlier than this utterance.

(9) Jane is relieved that the end of the race is earlier than noon, October 20, 2012.

But we could try to fill in Prior's argument. It seems that while (1) is true in the situation at the time it is said, (8) and (9) are false. And if they are false then neither can be the truth-condition for (1). Furthermore, there do not seem to be any other tenseless and A-vocabulary-free sentences that are better placed to be truth-conditions for (1) than these.

Why think that (8) and (9) are false in the situation? Suppose that I say (8) after watching Jane finish the race and seeing the look on her face. But she is out of earshot and does not hear me say it. Since she does not know that I uttered (8) she cannot be relieved that the end of the race was earlier than my utterance. Suppose also that Jane lost her watch and that there are no clocks around. Then since she does not know that the race ended before noon she cannot be relieved that it did.

Now relief is a propositional attitude. There are no "purely phenomenal" episodes of relief; one is always relieved that such-and-such is the case.[23] And attributions of propositions attitudes raise well-known philosophical puzzles. Take (9). Is it meant to be an attribution of a *de dicto* attitude or a *de re* attitude? If we are after a truth-condition for (1) then our best bet is to take (9) to be *de re*. Then (9) can be put more perspicuously like this:

(10) Noon, October 20, 2012 is such that Jane is relieved that the end of the race is earlier than it.

Is it so obvious that (10) is false? Jane has lost her watch and there are no clocks around. Certainly she does not know that "October 20, 2012" denotes a time after the end of the race. But does the absence of clocks prevent Jane from having *de re* attitudes about that time? She can still refer to it as "this time."

---

[23] It is the propositional attitude aspect of relief, not the phenomenal aspect, that is important in Prior's argument. That is why the argument is not an argument from experience.

Philosophers disagree about what it takes to have *de re* propositional attitudes. But I suspect that there is a defensible theory on which Jane can have the relevant *de re* attitudes.

I do not want to suggest that (10) is the only live contender for a tenseless and A-vocabulary-free truth-condition for (1). Another one—probably a better one—could come from the theory Robert Stalnaker defends in his paper "Indexical Belief."[24] Still, it seems to me that most block universe theorists concede that (1) lacks a tenseless and A-vocabulary-free truth-condition. They have another response to Prior's argument, and it is that response that I want to discuss.

This second response is easiest to understand if we reformulate the argument and use proposition-talk explicitly. Start with the thesis that propositions are sets of possible worlds. In the block universe theory we can think of possible worlds as complete, consistent, tenseless, and A-vocabulary-free descriptions of spacetime and its contents. Or we can think of them as the pictures I have been drawing to illustrate the block universe theory.

I, like everyone, have views about what the universe is like. I believe that there are black swans but I do not believe that there are any talking donkeys. There are lots of things that I believe and disbelieve. But my "total belief state" can be characterized by a single proposition, a single set of possible worlds. A possible world is in this set if and only if nothing I believe is incompatible with the hypothesis that that world is the actual world.

What goes for belief goes for other propositional attitudes, like relief. My "total relief state" can also be characterized by a single set of possible worlds. I am, so to speak, relieved that the actual world is somewhere inside that set.

Now Jane in our scenario is relieved that the race is in the past. What set of worlds characterizes her total relief state, if the block universe theory is true? Could it contain just the possible worlds according to which Jane finishes the race at noon, October 20, 2012? No, the argument goes, because she does not believe, and so is not relieved that, the race finished at that time. Similarly, no other set of block-universe possible worlds characterizes her total relief state. (I discussed a possible response to this claim in the context of the first formulation of the argument, so I will not dwell on it here.)

The argument continues: where the block universe seems to fail MST-Supertense succeeds.[25] In MST-Supertense we can think of a possible world as a

---

[24] Suppose that after she crosses the finish line but before the relief washes over her Jane says, or thinks to herself, the sentence "The race is over." Let S be this sentence token. Then, roughly speaking, the Stalnakerian truth-condition for (1) is that Jane is relieved that S expresses a truth. But this does not do full justice to the theory.

[25] Prior of course was a presentist, but his argument supports any theory that rejects reductionism about tense.

complete, consistent, superpresent-tensed description of spacetime and its contents. Each one will include a statement saying which time super-is present, as well as descriptions of what spacetime super-will be like and super-was like. Or we can think of possible worlds in MST-Supertense as the sequences of pictures I have drawn to illustrate the theory.

There is a set of these possible worlds that looks to adequately characterize Jane's relief state. That set includes just those possible worlds in which the end of the race is earlier than the time that is present. Figure 12.9 depicts a few of the worlds in Jane's relief state. While the set characterizing Jane's relief state contains some worlds in which the race ends at noon, it contains others in which it ends at some other time. So her having that relief state is compatible with her having no opinion at all about what time the race finished.

That is the more souped-up version of Prior's argument. The standard block universe response to it is to deny the premise

(11) Each person's relief-state can be characterized by a set of possible worlds.

Instead, the response continues, relief states can be characterized using things similar to possible worlds, namely centered possible worlds.[26] A centered possible world is a triple (w, t, p) of a (block universe) possible world w, a time t, and a person p. The existence of centered worlds is certainly compatible with the block universe theory of time. And Jane's relief state can be characterized by the set

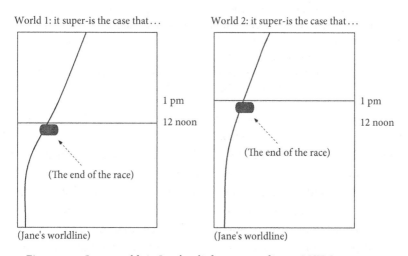

Figure 12.9 Some worlds in Jane's relief state according to MST-Supertense.

[26] Sider discusses the idea of using things like centered worlds to respond to Prior's argument in some detail in *Four-Dimensionalism* (pp. 18–21). The idea of using centered worlds to characterize belief states has been widely-discussed; perhaps the most popular framework is one Lewis describes in "Attitudes *De Dicto* and *De Se*."

containing just those centered worlds (w, t, p) in which p is Jane, Jane runs the race in w, and the race ends before t. So defenders of the block universe theory do have a way to characterize Jane's relief state, if they reject (11).

This solution is not very good without a story about what one is doing when one characterizes someone's relief, or belief, or other attitude state with a set of centered worlds. Interestingly, the standard story only really makes sense if the stage theory is true. Let me explain why.

The standard story is David Lewis's. Here is how he puts it:

What happens when [someone] has a propositional attitude, rightly so called? Take belief. What happens when he believes a proposition, say the proposition that cyanoacrylate glue dissolves in acetone?

Answer: he locates himself in a region of logical space. There are worlds where cyanoacrylate dissolves in acetone and worlds where it doesn't. He has a belief about himself: namely, that he inhabits one of the worlds where it does. . . .

A proposition divides the populace. Some are privileged to inhabit worlds where cyanoacrylate dissolves in acetone, others are not. . . . Someone who believes a proposition, and thereby locates himself in logical space, also places himself within the divided population. He has a partial opinion as to who he is. He is one of this class, not one of that class. To believe a proposition is to identify oneself as a member of a subpopulation comprising the inhabitants of the region of logical space where the proposition holds. Note that the boundaries of such a subpopulation follow the borders between world and world. Either all the inhabitants of a world belong, or none do. . . .

. . . [But] there is something arbitrary about taking the objects of belief always as sets of worlds. We are scattered not only through logical space, but also through ordinary space and time. We can have beliefs whereby we locate ourselves in logical space. Why not also beliefs whereby we locate ourselves in ordinary time and space? . . . We can identify ourselves as members of subpopulations whose boundaries follow the borders of the worlds. Why not also as members of subpopulations whose boundaries don't follow the borders of the worlds? (Attitudes *De Dicto* and *De Se*, p. 137.)

This is an argument, which does not mention tense explicitly, that sets of possible worlds are inadequate for characterizing belief states. It is worth asking whether the argument is any good. But this is not the place to criticize it.[27] Instead I want to look at Lewis's replacement for sets of possible worlds. He uses sets of possible individuals, and says that to characterize someone's belief state with a set of possible individuals is to attribute to him the belief that he is one of the individuals in that set. Now a set of possible individuals determines a set of (world, individual) pairs. So this passage from Lewis can be read as both explaining the

---

[27] Robert Stalnaker defends the claim that possible worlds are adequate against Lewis's argument in *Our Knowledge of the Internal World*. More recently Herman Cappelen and Josh Dever attack Lewis's conclusion that centered worlds are needed to characterize belief states in *The Inessential Indexical*.

need for, and helping to make intelligible, characterizing belief states using sets of centered worlds, where those centered worlds are world-individual pairs.

But this is not the kind of centered world we needed to respond to Prior's argument. There we needed world-time-individual centered worlds, not world-individual centered worlds.

This point bears repeating from a different angle. On Lewis's story, to believe is to locate oneself in "logical space," to have an opinion about which possible world one is in, and also to locate oneself in space and time within one or more possible worlds. What goes for belief also goes for relief: on Lewis's story one can be relieved not just about which possible world one is in, but also about where in space and time within a given possible world one is. This story seems to help with Prior's argument. We can say that Jane's relief is a self-locating attitude; she is relieved that she is located at a time after the end of the race, rather than at a time during or before the race.

But in fact when you think about it this story does not help at all. Where in time is Jane located? Talk of where in time Jane is located is equivalent to talk of which times Jane exists at. And we usually think that she exists at lots of times, all the times between the time of her birth and the time of her death. These certainly include times during the race. If that is right then she cannot be relieved that she is located at a time after the end of the race, unless she is deluded.[28]

Defenders of the block universe theory often accuse Prior's argument of being in bad company.[29] "If Prior's argument shows that the block universe is false because it leaves out facts about which time is present," they will say, "then a perfectly analogous argument shows that the block universe is false because it leaves out facts about which person is me." I think this is wrong. Consider Lewis's example of the two gods in "Attitudes *De Dicto* and *De Se*." There are two gods and each knows exactly which "block universe possible world" is actual. Each knows that there are exactly two gods, that one lives on the top of the highest mountain, and the other lives at the bottom of the deepest valley. But, says Lewis, there is still something they do not know. Neither knows whether he is the god who lives on the highest mountain or the one who lives in the deepest valley. (Again, there is dispute about whether Lewis is right to say this, but set that aside.) Now one might go on to argue that since there is still something the gods do not know, the block universe theory is false. (This is not Lewis's argument!) The true theory says that there is an "objectively distinguished individual" (just as

---

[28] Craig also expresses skepticism that standard treatments of self-locating belief and ignorance in the block universe theory make sense of belief and ignorance about the time (*The Tensed Theory of Time*, p. 128).

[29] See, for example, Sider, *Four-Dimensionalism* (p. 19), and Callender, "The Common Now" (p. 343).

MST-Supertense says that there is an objectively distinguished time). What the gods do not know, on this view, is whether it is the god on the mountain or the god on the valley who is objectively distinguished.

Lewis's account of what job centered worlds do does provide a reply to this argument. The defender of the block universe can say, what the gods lack is not knowledge of which possible world is actual, but of who in that possible world they are. We can characterize their ignorance using world-individual pairs rather than possible worlds. But motivating this response to this argument does not automatically motivate the centered-worlds response to Prior's. For knowing who one is does not suffice for knowing what time it is.

If you like Lewis's story there is a way around this problem. Accept the stage theory of persistence.[30] Then Jane exists at just one time, a time after the end of the race. So she does not need to be deluded to be relieved that she is located at that time. More generally, if the stage theory is true, and each individual exists at just one time, then a world-individual pair determines a world-time-individual triple. The time is just the time at which that individual exists. So the kind of centered world that Lewis's story motivates can do the work needed to answer Prior's argument.[31]

While this is interesting I do not think it is a strong argument for the stage theory. Those who like the centered worlds response to Prior's argument but do not like the stage theory can just tell a different story about what one is doing when one characterizes someone's attitude state using a set of centered worlds. They can say that "self-locating information" is a bad name for sets of centered worlds and leads to a misleading description of the job-centered worlds are supposed to do. When my belief state is characterized by the triple (the actual world; noon, October 20, 2012; Brad Skow) I am not locating myself in time—for I am located at many times. What I am doing is something like locating my current experiences, or my current act of awareness, or (the perdurance theorist may say) my current time-slice, in time.

[30] Cappelen and Hawthorne, in *Relativism and Monadic Truth* (p. 52), offer the fact that Lewis's story requires the stage theory to be one reason (among many they discuss) to reject the use of centered worlds to model belief states.

[31] David Lewis wrote in one place that "some cases of belief *de se* [i.e., self-locating belief] can be better understood if we take the believer not as a continuant but as a more-or-less momentary time-slice thereof" ("Attitudes *De Dicto* and *De Se*, p. 143). On the next page he says that when someone wonders what time it is, "it is time-slices of him that do the wondering." But it is *I* who wonder and then come to have beliefs about what time it is. And the perdurance theory, which Lewis endorses, denies that I am a time-slice. Lewis's remarks here seem to me to mark a slide from the perdurance theory to the stage theory. There are other places where Lewis's use of the perdurance theory to solve philosophical puzzles shows a tendency to think in stage-theoretic terms. See for example Sider's discussion of Lewis's discussion of fission, personal identity, and "what matters" (*Four-Dimensionalism*, p. 202).

13

Concluding Thoughts: Passage and Time-biased Preferences

Here ends my defense of the block universe theory's account of the passage of time. There is one more topic worth talking about, but I am afraid I can only introduce it. I am unable to say anything enlightening about it.

At the beginning of Chapter 11 I said that the usual instruction given to someone looking for reasons to believe in robust passage is: attend to your experience! But arguments from experience are not the only kind of argument for objective becoming. As I said in an appendix to Chapter 12, Prior's "Thank Goodness That's Over" argument is not an argument from experience. Now discussions of Prior's argument often wander away from the argument Prior actually gave (and not toward an argument that is an argument from experience). The question shifts from

> Is the truth of "Jane is relieved that the race is over" consistent with reductionism about tense?

to

> Why is Jane relieved after the race is over but not before?

to

> Is it rational for Jane to be relieved only after the race is over?

Although in asking these questions we have officially left Prior's argument behind, we are heading somewhere interesting.[1]

The question why Jane is relieved only after the race is not actually the best one to focus on, because it is easily answered from within the block universe theory. Jane is relieved after the race is over because she has been looking forward to its

---

[1] The shift appears in chapter 3 of Mellor's *Real Time,* and the influence of that book and its successor *Real Time II* may have encouraged the attribution of these questions to Prior.

being over; plus, she couldn't be relieved that the race is over before the race is over, unless she were deeply confused; what she could be is relieved that the race is about to start, or that the race will be over; and it is easy to imagine the story so that she is or is not relieved about these things, and for perfectly good reasons.

The second question, the one about whether Jane is being rational—that one is worth focusing on. And generalizing. Derek Parfit pointed out that when it comes to pleasure and pain we are biased toward the future (*Reasons and Persons*, §64). Suppose I wake in the hospital and cannot remember the last few days. A nurse comes in and tells me that her paperwork is mixed up, all she knows is that either I am the patient who is waking up the day after ten hours of painful surgery (during which I was wide awake), or I am the patient who is about to go in for one hour of painful surgery (during which I will be wide awake, and after which I will be given a drug that puts me to sleep and makes me forget the operation). When she leaves to sort out who I am I think to myself (and every one of you would think the same): I hope I am the patient with the operation in the past! Suppose on the other hand that I learn that either I did eat ten chocolate bars yesterday (and then forgot), or that I will eat one chocolate bar today (and will then forget). Now I hope that I am the person with the chocolate in the future. That is what it is to be biased toward the future: to prefer that pains be in the past and pleasures be in the future.

We certainly are biased toward the future. Is it rational to be biased toward the future?[2] Some philosophers think the answer is a resounding yes and argue that this spells trouble for the block universe theory. For, they say, if the block universe theory is true then the bias toward the future is not rational.[3]

I do not see how this argument is supposed to work. One way to buttress the argument is to use the notion of welfare, or well-being. Talk about well-being is talk about what makes a life good for the one who lives it, or makes life worth living. One theory, welfare hedonism, says that one life is better than another if and only if the total amount of pleasure in the first life, minus the total amount of pain in the first life, is greater than the corresponding quantity in the second life. If welfare hedonism is true then a reason for thinking that the bias toward the future is irrational is that people exhibiting it prefer a worse life to a better

---

[2] As with Jane's relief, we might also ask why we are biased toward the future. Horwich, in *Asymmetries in Time* (p. 197), suggested that being biased is a fitter trait than being neutral, so the action of natural selection explains why we are biased. But this does not help with the question of rationality. Adaptive beliefs and behaviors can easily be irrational.

[3] Schlesinger leans on something like this argument, though he is focused on the rationality of things like relief and anxiety, not of preferences. The quote from him below is his statement of one of the premises.

life. In the hospital I prefer to be the patient who experiences more pain—so long as that pain is in the past. And I prefer to be the person who eats less chocolate (and chocolate is a great pleasure)—so long as that chocolate is in the future.

But it is not in general irrational to prefer a worse life to a better. People sacrifice themselves all the time—for their children, their job, their country—and in so doing make things worse for themselves in order that things be better for others. And by doing this they may be doing what they most prefer to do. But they are not irrational.

This consideration only goes so far. Preferring that the operation be in the past does not feel like a sacrifice. So here is another point. I agree with Parfit that it is not obvious that the existence of objective becoming would help. Why would the bias toward the future be rational if there were robust passage? It is certainly true that if I have fewer pains in the future then there are fewer pains that I have yet to experience. But, this is true in the block universe theory too, so being able to say it is no advantage of the moving spotlight theory (the block universe theory does not deny that tensed sentences are true, only that they are fundamental truths).

A fan of this argument might say that welfare hedonism is the wrong theory of welfare to be working with. Instead we should focus on a new theory, tensed welfare hedonism. This theory presupposes the moving spotlight theory. It says that a life is better to the extent that the total amount of future pleasure, minus the total amount of future pain, is greater. If the moving spotlight theory is true, and tensed welfare hedonism is true, then in the hospital I do not prefer the worse life to the better one. I straightforwardly prefer the better one.

Here, finally, we have a combination of views that includes that moving spotlight theory and explains, if true, why the bias toward the future is rational. But tensed welfare hedonism is weird. It implies that my son's life super-is much better than mine (since he has so much more of his to live—I am assuming that no terrible calamities await him) but that it super-will be the case (in 200 superyears) that both our lives are equally worthless.

Tensed welfare hedonism is also mysterious. Why should it be that future and present pains are bad but past ones are not? The theory has no answer.[4]

Maybe we could learn to live with these aspects of the theory. But still, tensed welfare hedonism cannot do the job it is here recruited to do, namely make our preferences rational. For while we have time-biased preferences about our own

---

[4] One of Zimmerman's complaints about the moving spotlight theory can be read as an objection to the use of tensed welfare hedonism to defend the rationality of our time-biased preferences: "If a pain is just as intrinsically painful when it has the spotlight upon it [as when it does not] ... why should the passage of the spotlight ... change our attitude toward it" ("The Privileged Present," p. 214)?

pleasures and pains we (sometimes) lack them about other people's. Suppose that the story about the patient in this hospital is not a story about me but is a story that I know to be true of my brother in Seattle. Then I prefer that he have the shorter operation, the one with the pain in the future.[5] But if tensed welfare hedonism is true then I am preferring that he have a worse life.

I do not think that the fact that we are biased toward the future provides much of an argument for objective becoming. But I do think an interesting question emerges from reflecting on the argument. One of the argument's premises was: the bias toward the future is rational. Is it? Schlesinger wrote about it and related phenomena that "Nobody denounces these differences in attitude as irrational; nobody advocates that our attitudes are to be reformed in the light of a clear-headed analysis of temporal relations" (Aspects of Time, p. 34). But four years later Parfit suggested just that.[6] Suppose the block universe theory is true. Then maybe, if we could abandon the bias toward the future—with respect to preferences, and also with respect to other attitudes and emotions, like relief and regret—we should. Perhaps abandoning it would make our lives better in some ways. Perhaps if we were unbiased we would not, as we age, lament the fact that more and more of our lives are behind us.

Perhaps—or perhaps not. All that being unbiased requires is that we not discriminate between pleasures based on whether they are to the future or to the past. But there is more than one way to eliminate the bias. We could treat future pleasures the way we already treat past ones, or we could do it the other way around. If we eliminate the bias by continuing to be glad when a pleasure becomes past, won't we also continue to be upset when a pain becomes past? Then being unbiased will in some ways be worse. Grief will never fade.

I suppose we could pick and choose, and always be glad about the good things and never be upset about the bad ones.

I am asking whether our time-biased pattern of caring and valuing can find a home in the block universe, or if, having been convinced that the block universe is true, we should abandon it and adopt a different pattern. I hoped when I started this book to make progress on this difficult question. But I have not.

---

[5] Parfit mentions that we are not always time-biased towards others (Reasons and Persons, §69). Whether our preferences about other people are time-biased depends on how far away they are from us ("far away" is not just a matter of spatial distance; factors such as how easily we could speak to them on the phone also matter). But all that matters for my argument is that in some circumstances our preferences are unbiased. (Hare, in "A Puzzle about Other-Directed Time-Bias," argues that it is irrational to have preferences that are time-biased when the person is far away but unbiased when the person is nearby.)

[6] It seems that Spinoza suggested it too, several centuries ago; see Cockburn, "Tense and Emotion."

References

Arntzenius, Frank. "Causal Paradoxes in Special Relativity." *British Journal for the Philosophy of Science* 41: 1990, 223–43.

Baker, Lynne Rudder. "Temporal Becoming: The Argument from Physics." *Philosophical Forum* 6: 1975, 218–36.

Balashov, Yuri. "Times of Our Lives: Negotiating the Presence of Experience." *American Philosophical Quarterly* 42: 2005, 295–309.

Barbour, Julian. *The End of Time.* Oxford University Press: 1999.

Barenblatt, G. I. *Scaling, Self-similarity, and Intermediate Asymptotics.* Cambridge University Press: 1996.

Barnes, Elizabeth and Ross P. Cameron. "Back to the Open Future." *Philosophical Perspectives* 25: 2011, 1–26.

Bayne, Tim. *The Unity of Consciousness.* Oxford University Press: 2010.

Belot, Gordon. "Dust, Time, and Symmetry." *British Journal for the Philosophy of Science* 56: 2005, 255–91.

——. "Symmetry and Equivalence." In Robert Batterman (ed.), *The Oxford Handbook of Philosophy of Physics.* Oxford University Press: 2013, 318–39.

Bigelow, John. "Worlds Enough for Time." *Nous* 25: 1991, 1–19.

Bourne, Craig. *A Future for Presentism.* Oxford University Press: 2006.

Broad, C. D. *Examination of McTaggart's Philosophy*, volume 2, part 1. Cambridge University Press: 1938.

——. *Scientific Thought.* Harcourt, Brace, and Company: 1923.

Brogaard, Berit. *Transient Truths: An Essay in the Metaphysics of Propositions.* Oxford University Press: 2012.

Burgess, John P. *Philosophical Logic.* Princeton University Press: 2009.

Byrne, Alex. "Experience and Content." *Philosophical Quarterly* 59: 2009, 429–51.

——. "Knowing What I See." In Declan Smithies and Daniel Stoljar (eds.), *Introspection and Consciousness.* Oxford University Press: 2012, 183–210.

——. "Sensory Qualities, Sensible Qualities, Sensational Qualities." In Brian P. McLaughlin and Ansgar Beckermann (eds.), *The Oxford Handbook of Philosophy of Mind.* Oxford University Press: 2009, 268–80.

Callender, Craig. "On Finding 'Real' Time in Quantum Mechanics." In William Lane Craig and Quentin Smith (eds.), *Absolute Simultaneity.* Routledge: 2008, 50–72.

——. Review of Mauro Dorato, *Time and Reality: Spacetime Physics and the Objectivity of Temporal Becoming.* British Journal for the Philosophy of Science 48: 1997, 117–20.

——. "Shedding Light on Time." *Philosophy of Science* 67: 2000, S587–99.

——. "The Common Now." *Philosophical Issues* 18: 2008, 339–61.

——. "Time's Ontic Voltage." In Adrian Bardon (ed.), *The Future of the Philosophy of Time.* Routledge: 2012, 73–98.

Cappelen, Herman, and Josh Dever. *The Inessential Indexical.* Oxford University Press: 2014.

Cappelen, Herman, and John Hawthorne. *Relativism and Monadic Truth.* Oxford University Press: 2009.

Clifton, Rob, and Mark Hogarth. "The Definability of Objective Becoming in Minkowski Spacetime." *Synthese* 103: 1995, 355–87.

Cockburn, David. "Tense and Emotion." In Robin Le Poidevin (ed.), *Questions of Time and Tense.* Oxford University Press: 1998, 77–91.

Craig, William Lane. *The Tensed Theory of Time: A Critical Examination.* Kluwer Academic Publishers: 2000.

——. *The Tenseless Theory of Time: A Critical Examination.* Kluwer Academic Publishers: 2000.

Craig, William Lane, and Quentin Smith (eds.). *Einstein, Relativity and Absolute Simultaneity.* Routledge: 2008.

Dainton, Barry. *Time and Space.* McGill-Queen's University Press: 2010.

——. "Time, Passage, and Immediate Experience." In Craig Callender (ed.), *The Oxford Handbook of Philosophy of Time.* Oxford University Press: 2011, 382–419.

Davies, P. *About Time: Einstein's Unfinished Revolution.* Penguin: 1995.

Dieks, Dennis. "Becoming, relativity and locality." In Dennis Dieks (ed.), *The Ontology of Spacetime.* Elsevier: 2006, 157–76.

Dorato, Mauro. "Absolute becoming, relational becoming and the arrow of time: Some non-conventional remarks on the relationship between physics and metaphysics." *Studies in History and Philosophy of Modern Physics* 37: 2006, 559–76.

——. *Time and Reality: Spacetime Physics and the Objectivity of Temporal Becoming.* CLUEB Press: 1995.

Dummett, Michael. "A Defense of McTagart's Proof of the Unreality of Time." *The Philosophical Review* 69: 1960, 497–504.

Earman, John. "Space-Time, or How to Solve Philosophical Problems and Dissolve Philosophical Muddles without Really Trying." *Journal of Philosophy* 67: 1970, 259–77.

——. *World Enough and Space-Time.* MIT Press: 1989.

Eddington, A. S. *Space, Time, and Gravitation.* Cambridge University Press: 1920.

Edwards, Harold M. *Advanced Calculus: A Differential Forms Approach.* Birkhaüser: 1994.

Evans, Gareth. *The Varieties of Reference.* Oxford University Press: 1982.

Falk, Arthur. "Time Plus the Whoosh and Whiz." In Aleksandar Jokić and Quentin Smith (eds.), *Time, Tense, and Reference.* The MIT Press: 2003, 211–50.

Ferré, Frederick. Grünbaum on Temporal Becoming: A Critique." *International Philosophical Quarterly* 12: 1972, 426–45.

——. "Grünbaum vs. Dobbs: The Need for Physical Transiency." *British Journal for the Philosophy of Science* 21: 1970, 270–80.

Field, Hartry. "No Fact of the Matter." *Australasian Journal of Philosophy* 81: 2003, 457–80.

——. *Science Without Numbers.* Princeton University Press: 1980.

Fine, Kit. *Modality and Tense: Philosophical Papers.* Oxford University Press: 2005.

——. "Tense and Reality." In *Modality and Tense.* 261–320.

Forrest, Peter. "Relativity, The Passage of Time, and the Cosmic Clock." In Dennis Dieks (ed.), *The Ontology of Spacetime II.* Elsevier: 2008, 245–54.

——. "The real but dead past: a reply to Braddon-Mitchell." *Analysis* 64: 2004, 358–62.

Geroch, Robert. *General Relativity from A to B.* University of Chicago Press: 1978.

Gibson, Ian, and Oliver Pooley. "Relativistic Persistence." *Philosophical Perspectives* 20: 2006, 157–98.

Gilmore, Cody. "Persistence and Location in Relativistic Spacetime." *Philosophical Compass* 3: 2008, 1224–54.

——. "Where in the Relativistic World Are We?" *Philosophical Perspectives* 20: 2006, 199–236.

Giulini, Domenico. "Uniqueness of Simultaneity." *British Journal for the Philosophy of Science* 52: 2001, 651–70.

Gödel, Kurt. "A Remark about the relationship between relativity theory and idealistic philosophy." In Solomon Feferman et al. (eds.), *Collected Works,* volume II. Oxford University Press: 1990, 202–7.

Hare, Caspar. "A Puzzle About Other-Directed Time-Bias." *Australasian Journal of Philosophy* 86: 2008, 267–77.

——. *On Myself, and Other, Less Important, Subjects.* Princeton University Press: 2009.

Haslanger, Sally. "Endurance and Temporary Intrinsics." *Analysis* 49: 1989, 119–25.

Hawley, Katherine. *How Things Persist.* Oxford University Press: 2001.

Hestevold, H. Scott. "Passage and the Presence of Experience." *Philosophy and Phenomenological Research* 50: 1990, 537–52.

Hinchliff, Mark. "A Defense of Presentism in a Relativistic Setting." *Philosophy of Science* 67: 2000, S575–86.

Horwich, Paul. *Asymmetries in Time.* MIT Press: 1987.

Huddleston, Rodney, and Geoffrey K. Pullum. *A Student's Introduction to English Grammar.* Cambridge University Press: 2005.

Jeans, James. "Man and Universe." In J. Jeans, W. Bragg, E. Appleton, E. Mellenby, J. Haldane, and J. Huxley (eds.), *Scientific Progress.* The Macmillan Company: 1936, 11–38.

Kahneman, Daniel. *Thinking, Fast and Slow.* Farrar, Straus and Giroux: 2011.

Kaplan, David. "Demonstratives." In Joseph Almog, John Perry, and Howard Wettstein (eds.), *Themes from Kaplan.* Oxford University Press: 1989, 481–563.

King, Jeffrey. "Tense, Modality, and Semantic Values." *Philosophical Perspectives* 17: 2003, 195–245.

Le Poidevin, Robin. "Relative Realities." *Studies in History and Philosophy of Modern Physics* 28: 1997, 541–6.

——. *The Images of Time.* Oxford University Press: 2007.

Lewis, David. "Attitudes De Dicto and De Se." In *Philosophical Papers,* volume 1. Oxford University Press: 1983, 133–56.

——. *On the Plurality of Worlds.* Blackwell: 1986.

——. "Rearrangement of paticles: reply to Lowe." *Analysis* 48: 1988, 65–72.

Lockwood, Michael. *Mind, Brain, and the Quantum.* Blackwell: 1989.

Ludlow, Peter. *Semantics, Tense, and Time.* MIT Press: 1999.

McCall, Storrs. "Objective Time Flow." *Philosophy of Science* 43: 1976, 337–62.

MacFarlane, John. "Truth in the Garden of Forking Paths." In Max Kölbel and Manuel García-Carpintero (eds.), *Relative Truth.* Oxford University Press: 2008, 81–102.

McTaggart, J. M. E. *The Nature of Existence,* volume 2. Cambridge University Press: 1927.

——. "The Unreality of Time." *Mind* 17: 1908, 457–74.

Malament, David. "Causal Theories of Time and the Conventionality of Simultaneity." *Nous* 11: 1977, 293–300.

Markosian, Ned. "A Defense of Presentism." In Dean Zimmerman (ed.), *Oxford Studies in Metaphysics,* volume 1. Oxford University Press: 2004, 47–82.

——. "How Fast Does Time Pass?" *Philosophy and Phenomenological Research* 53: 1993, 829–44.

Maudlin, Tim. "On the Passing of Time." In *The Metaphysics Within Physics.* Oxford University Press: 2007, 104–42.

——. *Quantum Non-Locality and Relativity: Metaphysical Intimations of Modern Physics.* 3rd edition. Oxford University Press: 2011.

Mellor, D. H. *Real Time.* Cambridge University Press: 1981.

——. *Real Time II.* Routledge: 1998.

Mermin, N. David. *It's About Time: Understanding Einstein's Relativity.* Princeton University Press: 2005.

Merricks, Trenton. Review of Katherine Hawley, *How Things Persist. Mind* 112: 2003, 146–8.

Minkowski, Hermann. "Space and Time." In *The Principle of Relativity.* Dover: 1952.

Monton, Bradley. "Presentism and Quantum Gravity." In Dennis Dieks (ed.), *The Ontology of Spacetime.* Elsevier: 2006, 263–80.

Oaklander, L. Nathan. "On the Experience of Tenseless Time." In Oaklander and Smith (eds.), *The New Theory of Time,* 344–50.

Oaklander, L. Nathan, and Quentin Smith (eds.). *The New Theory of Time.* Yale University Press: 1994.

Olson, Eric T. "The Rate of Time's Passage." *Analysis* 61: 2009, 3–9.

Palacios, J. *Dimensional Analysis,* 2nd edition. St. Martin's Press: 1964.

Parfit, Derek. *Reasons and Persons.* Oxford University Press: 1984.

Pariyadath, Vani, and David Eagleman. "The Effect of Predictability on Subjective Duration." *PLoS ONE* 2: 2007, e1264.

Parsons, Josh. "A-Theory for B-Theorists." *The Philosophical Quarterly* 52: 2002, 1–20.

——. "Theories of Location." In Dean Zimmerman (ed.), *Oxford Studies in Metaphysics,* volume 3. Oxford University Press: 2007, 201–32.

Partee, Barbara. "Some Structural Analogies between Tense and Pronouns in English." *Journal of Philosophy* 70: 1973, 601–7.

Paul, L. A. "Temporal Experience." *The Journal of Philosophy* 107: 2010, 333–59.

——. "Truth Conditions of Tensed Sentence Types." *Synthese* 111: 1997, 53–71.

Percival, Philip. "A Presentist's Refutation of Mellor's McTaggart." *Royal Institute of Philosophy Supplement* 50: 2002, 91–118.

Phillips, Ian. "Perceiving the Passage of Time." *Proceedings of the Aristotelian Society* 113: 2013, 225–52.

Pooley, Olver. "Relativity, the Open Future, and the Passage of Time." *Proceedings of the Aristotelian Society* 113: 2013, 321–63.

Price, Huw. *Time's Arrow and Archimedes' Point.* Oxford University Press: 1996.

Priest, Graham. "Tense and Truth Conditions." *Analysis* 46: 1986, 162–6.

Prior, A. N. "Changes in Events and Changes in Things." In *Papers on Time and Tense*, 7–20.

——. *Papers on Time and Tense*, new edition. Per Hasle, Peter Ohrstrom, Torben Braüner, and Jack Copland (eds.). Oxford University Press: 2003.

——. *Past, Present and Future*. Oxford University Press: 1967.

——. "Some Free Thinking about Time." In B. J. Copeland (ed.), *Logic and Reality: Essays on the Legacy of Arthur Prior*. Oxford University Press: 1996, 47–51.

——. "Tense Logic and the Logic and Earlier and Later." In *Papers on Time and Tense*, 117–38.

——. "Thank Goodness That's Over." *Philosophy* 34: 1959, 12–17.

——. "The Notion of the Present." *Studium Generale* 23: 1970, 245–48.

——. "Time After Time." *Mind* 67: 1958, 244–6.

Prosser, Simon. "Could We Experience The Passage of Time?" *Ratio* 20: 2007, 75–90.

——. "Passage and Perception." *Nous* 47: 2013, 69–84.

Putnam, Hilary. "Time and Physical Geometry." *The Journal of Philosophy* 64: 1967, 240–7.

Russell, Bertrand. "On the Experience of Time." *The Monist* 25: 1915, 212–33.

——. *The Principles of Mathematics*. Cambridge University Press: 1903.

Saunders, Simon. "How Relativity Contradicts Presentism." In Craig Callender (ed.), *Time, Reality and Experience*. Cambridge University Press: 2002, 277–92.

Savitt, Steven F. "A Limited Defense of Passage." *American Philosophical Quarterly* 38: 2001, 261–70.

——. "On Absolute Becoming and the Myth of Passage." In Craig Callender (ed.), *Time, Reality and Experience*. Cambridge University Press: 2002, 153–67.

Schlesinger, George. *Aspects of Time*, Hackett: 1980.

——. "How Time Flies." *Mind* 91: 1982, 501–23.

——. "The Stream of Time." In Oaklander and Smith (eds.), *The New Theory of Time*, 257–85.

Sider, Theodore. "All The World's A Stage." *Australasian Journal of Philosophy* 74: 1996, 433–53.

——. *Four-Dimensionalism*. Oxford University Press: 2001.

——. "Presentism and Ontological Commitment." *Journal of Philosophy* 96: 1999, 325–47.

——. *Writing the Book of the World*. Oxford University Press: 2011.

Siegel, Susanna. *The Contents of Visual Experience*. Oxford University Press: 2011.

Skow, Bradford. "Experience and the Passage of Time." *Philosophical Perspectives* 25: 2011, 359–87.

——. "On the Meaning of the Question 'How Fast Does Time Pass?'." *Philosophical Studies* 155: 2011, 325–44.

——. "Relativity and the Moving Spotlight." *The Journal of Philosophy* 106: 2009, 666–78.

——. "What Makes Time Different From Space?" *Nous* 41: 2007, 227–52.

Smart, J. J. C. *Philosophy and Scientific Realism*. Routledge: 1963.

——. "The River of Time." *Mind* 58: 483–94.

——. "Time and Becoming." In Peter van Inwagen (ed.), *Time and Cause: Essays presented to Richard Taylor*. D. Reidel: 1980, 3–15.

Smith, Nicholas J. J. "Inconsistency in the A-theory." *Philosophical Studies* 156: 2011, 231–47.

Smith, Quentin. "The Phenomenology of A-Time." In Oaklander and Smith (eds.), *The New Theory of Time*, 351–9.

Solomyak, Olla. "Actuality and the amodal perspective." *Philosophical Studies* 164: 2013, 15–40.

Spencer, Jack. "Relativity and Ontology." Doctoral Dissertation, Princeton University: 2013.

Sperry, R. W. "Lateral Specialization in the Surgically Separated Hemispheres." In F. O. Schmitt and F. G. Worden (eds.), *Neuroscience, 3rd Study Programme.* MIT Press: 1974, 5–19.

Spivak, Michael. *Calculus on Manifolds.* Westview Press: 1971.

Stalnaker, Robert. "Indexical Belief." In *Context and Content.* Oxford University Press: 1999, 130–49.

——. *Our Knowledge of the Internal World.* Oxford University Press: 2008.

Stein, Howard. "On Einstein–Minkowski Space-Time." *The Journal of Philosophy* 65: 1968, 5–23.

——. "On Relativity Theory and the Openness of the Future." *Philosophy of Science* 58: 1991, 147–67.

Stetson, C., Fiesta, M. P., and Eagleman, D. M. "Does time really slow down during a frightening event?" *PLoS One*: 2007, 1295.

Stevens, S. S. "On the Theory of Scales of Measurement." *Science* 103: 1946, 677–80.

Szabó, Zoltán. "Counting Across Times." *Philosophical Perspectives* 20: 2007, 399–426.

Taylor, Richard. *Metaphysics*, 4th edition. Prentice Hall: 1992.

Thomas, Dylan. *Under Mik Wood.* New Directions: 1954.

Tooley, Michael. *Time, Tense, and Causation.* Oxford University Press: 1997.

Torre, Stephan. "Truth-conditions, truth-bearers, and the New B-theory of time." *Philosophical Studies* 142: 2009, 325–44.

Van Cleve, James. "If Meinong Is Wrong, Is McTaggart Right?" *Philosophical Topics* 24: 1996, 231–54.

——. "Rates of Passage." *Analytic Philosophy* 52: 2011, 141–70.

Van Inwagen, Peter. *Metaphysics*, 3rd edition. Westview Press: 2009.

Weyl, Hermann. *The Philosophy of Mathematics and Natural Science.* Princeton University Press: 1949.

Williams, Donald C. "The Myth of Passage." *The Journal of Philosophy* 48: 1951, 457–72.

Williamson, Timothy. *Modal Logic and Metaphysics.* Oxford University Press: 2031.

——. "Necessity Existents." In A. O'hear (ed.), *Logic, Thought and Language.* Cambridge University Press: 2002, 233–51.

——. *Vagueness.* Routledge: 1994.

Wüthrich, Christian. "No Presentism in Quantum Gravity." In Vesselin Petkov (ed.), *Space, Time, and Spacetime: Physical and Philosophical Implications of Minkowski's Unification of Space and Time.* Springer: 2012, 257–78.

——. "The fate of presentism in modern physics." In Roberto Ciunti, Kristie Miller, and Giuliano Torrengo (eds.), *New Papers on the Present.* Philosophia Verlag: 2013, 91–131.

Yablo, Stephen. *Aboutness.* Princeton University Press, 2014.

Yalcin, Seth. "Semantics and Metasemantics in the Context of Generative Grammar." In A. Burgess and B. Sherman (eds.), *Metasemantics*. Oxford University Press: 2014, 17–54.

Zimmerman, Dean. "Presentism and the Space-Time Manifold." In Craig Callender (ed.), *The Oxford Handbook of Philosophy of Time*. Oxford University Press: 2011, 163–244.

——. "The A-theory of time, the B-theory of time, and 'Taking Tense Seriously'." *Dialectica* 59: 2005, 401–57.

——. "The Privileged Present: Defending an 'A-theory' of Time." In Theodore Sider, John Hawthorne, and Dean Zimmerman (eds.), *Contemporary Debates in Metaphysics*. Blackwell: 2008, 211–25.

Index

anemic passage of time, *see* passage of time
Arntzenius, Frank 134n
A series, the 49
A-theory of time 18n, 89
availability 208
    and the moving spotlight theory 214–16
    and the perdurance theory 224–25
    and the stage theory 219–21
    as time-independent 212–13
    as time-relative 211–12
    as place-relative 223–4
A-vocabulary 12
    and special relativity 176–7

Baker, Lynne Rudder 212n
Balashov, Yuri 210n, 212n, 219n
Barbour, Julian 110n
Barenblatt, G. I. 117n, 122n
Barnes, Elizabeth 77n
Bayne, Tim 224n
Belot, Gordon 148n, 168n, 172n, 174
bias toward the future, the 234–6
Bigelow, John 53n
block universe theory of time; *see also* passage
    of time; reductionism about tense
    and motion of time 44
    and perspective-talk 63, 64n, 90
    and redundant temporal modifiers 96
    and time-relative truth 13–14, 69, 91
    as lacking robust passage 17–18, 22–7
    contrasted with MST-Time 59
    contrasted with presentism 28, 32
    spatial analogue of 222
    statement of 4, 10–17
Bourne, Craig 28n, 32n, 171, 215n
branching time theory 71–2; *see also* future
    indeterminacy
Broad, C. D. 18, 71n
    against supertime 50–2
    on McTaggart's argument 83, 85–6, 88
    on the moving spotlight theory 46
    on the rate of time's passage 101–2, 103, 105,
        111–12, 114, 125
Brogaard, Berit 21n
B series, the 49
B-theory of time 2n, 18n
Burgess, John P. 106n
Byrne, Alex 190n, 194n, 201n

calculus 23n, 124
Callender, Craig 67n, 82n, 160, 161n, 167n, 171n,
    173n, 174n, 193n, 231n
Cameron, Ross P. 77n
Cappelen, Herman 230n, 232n
change; *see also* passage of time; rate of change
    anemic change 27, 50
    argument that the block universe lacks 24–7
    as necessary condition on robust passage 22,
        50, 67
    ordinary vs extraordinary 36–7
    perception of, as favoring robust
        passage 200–1
    quasi-change 38
    real change 27
    robust change 27, 33, 35, 50, 59n, 201, 213–14
    Russell's theory of 23
chocolate 192, 234–5
class of systems of scales, *see* scale of
    measurement
Clifton, Rob 171n
Cockburn, David 236n
content of experience 190–1
Craig, William Lane 148n, 173n, 212n, 231n

Dever, Josh 230n
Dainton, Barry 200n
Davies, P. 191n
Dieks, Dennis 18n
dimension of indicies, *see* time (A-time);
    supertime; time-like dimension
dimension function 119–20, 123
dimensionless number 126
Dorato, Mauro 18n, 171n
Dummett, Michael 67n, 87n

Eagleman, David 200n
Earman, John 9n, 148n
Eddington, A. S. 37
Edwards, Harold M. 124
endurance theory of persistence 182, 217, 225
    in tension with the block universe theory 181,
        183–6
    spatial analogue of 221
eternalism 10
Evans, Gareth 194n

events 1, 33, 41–2, 95–6, 100
experience, *see* content of experience;
    phenomenal character

Falk, Arthur 12n, 189, 191n
Ferré, Frederick 209n, 212n
Field, Hartry 113n
Fine, Kit 66n, 67–8
foliation by instants 154
Forrest, Peter 149n, 205n
fundamentality 13–15
fundamental quantity, *see* quantity
fundamental unit 117; *see also* scale of
    measurement; quantity
future
    as open 74–7
    definitions of 165
future indeterminacy 74; *see also* reductionism
        about tense
    and the branching time theory 76, 78–80
    and the moving spotlight theory 74–6
    and the growing block theory 74, 80
    argument against reductive analyses of 76–8
future tense 11n

general relativity 171–3, 175
Geroch, Robert 130n
Gibson, Ian 176n
Gilmore, Cody 183n
Giulini, Domenico 140n
Gödel, Kurt 172n
growing block theory 70–1, 81; *see also* future
    indeterminacy

Hare, Caspar 160n, 236n
Haslanger, Sally 67n
Hawley, Katherine 183n, 218n, 219n
Hawthorne, John 232n
hedonism, *see* welfare hedonism
Hestevold, H. Scott 191n, 193n, 195, 201, 210n
Hinchliff, Mark 156n, 161n, 167n
Hogarth, Rob 171n
Horwich, Paul 59n, 67n, 90n, 234n
Huddleston, Rodney 11n

illusion 190; *see also* passage of time
    cognitive illusion 191
indeterminacy, *see* future indeterminacy
instant, *see* time
intentionalism 202

Jeans, James 173n

Kahneman, Daniel 191n
Kaplan, David 20
King, Jefrey 30n

Leibniz notation 124
Le Poidevin, Robin 171n, 193n
Lewis, David 77, 81, 212n, 219n, 225n, 229n,
    230–2
location, spatiotemporal 179–81, 184
    in analysis of "wholly present" 183
    many meanings of 184–6
Lockwood, Michael 178, 179
Ludlow, Peter 16n

McCall, Storrs 37n, 72n
MacFarlane, John 69
McTaggart, J. M. E. 2n, 24, 36, 49, 86, 100, 101
    against the moving spotlight theory 83–5
Malament, David 140n
Markosian, Ned 28n, 32n, 105n, 127–9, 144n
Maudlin, Tim 37, 81n, 126n, 150n, 189n
measurable quantity, *see* quantity
Mellor, D. H. 83
    Mellor's reconstruction of McTaggart's
        argument 86–8
    Mellor's McTaggart-like argument 90–4
    on reductionism about tense 20
Mermin, N. David 130n
Merricks, Trenton 218n
metaphysically complete
    defined 26
    and presentism 34
    and the moving spotlight theory 48, 50, 53,
        60–1, 64–8, 85, 94, 96, 214, 223
metric supertense operators, *see* supertense
    operators
metric tense operators, *see* tense operators
Minkowski, Hermann 5
Minkowski spacetime, *see* spacetime
modal analyses of tense 53n
modal realism 77, 219n
Monton, Bradley 171n, 173n
moving spotlight theory, the; *see also* future
    indeterminacy; metaphysically complete;
    MST-Spacetime; MST-Superspacetime;
    MST-Supertense; MST-Supertime; MST-
    Time; passage of time; perspectivalism;
    rate of time's passage
    and the bias toward the future 235–6
    as containing two time-dimensions 54,
        88–90, 197
    deviant versions in Newtonian spacetime 158
    epistemic problem for, in relativity 157–9
    hybrid versions 213–16
    motion of time in 45, 47–9
    neo-classical versions, defined 159
    Schlesinger's version 51–2
    solipsism and 158–9
    spatial version analogue of 222–3
    type-N relativistic versions 146–8, 155–6, 172
    type-G relativistic versions 145–6, 173

motion through time; *see also* endurance theory of persistence; moving spotlight theory (hybrid versions); perdurance theory of persistence
  as metaphor for objective becoming 1, 36–7, 187
  in the block universe theory 180–1
  theory of 179–81, 187
MST-Spacetime 165–9, 173, 174–5
  and Howard Stein's theory 169–71
  and general relativity 172
MST-Superspacetime 161–2
MST-Supertense 52–4; *see also* supertense operators
  and McTaggart's argument 84–5
  and Mellor's argument 94
  and special relativity 162–5
  rate of time's passage in 105–9, 112
  truth-conditions for ordinary tense 54–8
MST-Supertime 45–7, 48–9
  rate of time's passage in 109–10
  truth-conditions for supertense 53
MST-Time 58–62, 114; *see also* perspectivalism
  and McTaggart's argument 84
  and Mellor's argument 92–4
  and Smith's argument 89
  and Van Cleve's argument 96
  rate of time's passage in 114
  truth-conditions for ordinary tense 62

Newtonian spacetime, *see* spacetime
nominalism 48

Oaklander, Nathan 20n, 210n, 225n
Oberlin College alumni, *see* Ferré, Frederick; Markosian, Ned; Sperry, R. W.
objective becoming, *see* branching time theory; growing block theory; moving spotlight theory; passage of time
Olson, Eric T. 115n
one second per second, *see* rate of change

Palacios, J. 112n
Parfit, Derek 234–6
Pariyadath, Vani 200n
Parsons, Josh 53n, 186n
Partee, Barbara 30n
passage of time; *see also* rate of time's passage
  anemic vs robust passage 2, 18, 27
  as an illusion 37, 189, 190, 194, 202n
  as change in which time is present 51–2
  as change of time, or reality, as a whole 49, 56, 61, 70–2, 187, 213–14
  as entailing changing truth-values 90
  as given in experience 188, 191–2
  as requiring (robust) change 22, 26, 50, 67

  in MST-Supertense, with no rate of passage 107–9
  in MST-Supertense, with a rate of passage 111
  in presentism 33, 38–41
  metaphors for 1, 36–7
past tense 30
past, definitions of 84–6, 165
Paul, L. A. 20n, 191n, 194
Percival, Philip 90n, 92n
perdurance theory of persistence 181–5, 217, 224–5, 232n; *see also* availability; reductionism about tense
perspectivalism 62–6, 90
perspective-independent fact 64, 66
perspectives, *see* superspacetime; supertime; time
phenomenal character 201
  as something that changes 205
phenomenal contrast, method of 199
Phillips, Ian 204
Pooley, Oliver 72n, 176
presence of experience, the 210n
present
  as boundary between past and future 165–6
  as phenomenal property 201–2
  as predicate of abstract times 39
  as visible property 192–9
  definitions of 74
  reductionism about 59
  spatial extent of 160, 169, 172, 174
presentism 2, 4n, 12, 27–30, 52; *see also* metaphysically complete; passage of time; reductionism about tense; tense operators
  and Mellor's argument 93n
  and special relativity 149, 156n, 169–70
  and robust change 27, 33
  analysis of change in 33, 38
  argument that it lacks robust passage 39–43
  serious presentism 42
Price, Huw 124–5, 189n
Priest, Graham 90n
Prior, A. N; *see also* presentism
  against the block universe theory 28
  against reductionism about tense 19, 225–7, 233
  as well-named 28n
  on change 38
  on instants of time 29n
  on McTaggart's argument 87
  on merely past individuals 39
  on special relativity 144n
  on the passage of time 33
  on the rate of time's passage 127
  Prior's version of presentism 30
propositions 20–1
  relative truth of 68–69, 91n
Prosser, Simon 189n, 195n, 203n

Pullum, Geoffrey K. 11n
Putnam, Hilary 149, 166n, 167n, 169–70, 171n

quantity 104, 115–16; *see also* dimension
    function, dimensionless number
  and rates of change 125–6
  broader and narrower notions of 116n
  dimensionless quantity 120–21, 124–6
  dimension of 118
  fundamental quantity 119, 121, 125
  supertemporal separation as 110
  unit for measuring 116–17
  values of a 116
quantum mechanics 76, 172–3

rate of change; *see also* quantity
  definition of 105, 126
  of one second per second 121, 125
  rateless change 104–5
  with respect to the "pure passage of
    time" 128–9
rate of time's passage; *see also* Broad, C. D.;
    MST-Supertense; MST-Time; Prior, A. N.
  in relativistic theories 156, 169
  measured in hours per miles run by
    Bikila 128
  perception of 200
reductionism about tense
  and future indeterminacy 78
  and presentism 29–30, 170
  and "thank goodness that's over" 226
  and the block universe theory 10–17
  and the moving spotlight theory 53, 54, 59,
    61, 85–6, 88
  and the perdurance theory 183
  and the stage theory 219
  reductionist truth-conditions 16
  reductionist truth-conditions in
    relativity 175–7
  versions of 19–21
robust passage of time, *see* passage of time
Russell, Bertrand 2n, 22–4, 26n

Saunders, Simon 149n, 150n, 161n
Savitt, Steven F. 18n, 84n
scale of measurement 116–18
  class of systems of scales 118–20
  system of scales/units 118
Schlesinger, George 51, 53n, 59n, 103, 112, 188,
    191, 234n, 236
Sider, Theodore 2n, 14n, 23n, 26, 30n, 32n, 42n,
    113n, 183n, 214n, 217n, 219n, 229n, 231n,
    232n
Siegel, Susanna 199
skepticism 189
Smart, J. J. C. 19, 36–7, 50, 51, 100, 102–3, 111–12,
    114, 189, 191n

Smith, Nicholas J. J. 89–90
Smith, Quentin 20n, 173n, 191n, 225n
Solomyak, Olla 68n
space 1, 4–9, 131, 164–5
spacelike line 150–2
spacetime 4–9, 70, 77
  foliation of 154, 172
  general relativistic spacetime 172n
  Newtonian spacetime 9, 130–1, 142, 143,
    150–1, 158
  Minkowski spacetime 130–9, 143, 144, 152–6
  spacetime points as perspectives on
    reality 165
spacetime interval 131–2
spatiotemporal location, *see* location,
    spatiotemporal
special relativity 140–1, 144, 149, 160, 171, 175
Spencer, Jack 67n
Sperry, R. W. 224
Spivak, Michael 124n
stage theory of persistence 217–21, 224–5; *see
    also* availability
  and centered possible worlds 230, 232
Stalnaker, Robert 228, 230n
Stein, Howard 149n, 160, 161n, 169–71
supersecond 110
  objections to fundamentality of 112–13
superspacetime 160–1; *see also*
    MST-Superspacetime
  superspacetimes as perspectives on
    reality 161
supertense operators 52
  truth-conditions for, *see* MST-Supertime
  metric operators 111
  metric operators defined using non-metric
    operators 113
  problems with, in relativity 162–5
supertime 178–79; *see also* MST-supertime
  distance in 109–10
  robust passage of 51
  supertimes as perspectives on reality 60
Stetson, C. 200n
Stevens, S. S. 116n
system of scales of measurement, *see* scale of
    measurement
system of units, *see* scale of measurement
Szabó, Zoltán 12n

Taylor, Richard 191n
temporal parts 62n, 182–3; *see also* perdurance
    theory of persistence
tenseless verb 11–12, 54
tense operators; *see also* supertense operators
  metric tense operators 38n, 109
  use in presentism 30
tensed welfare hedonism 235–6

"thank goodness that's over," *see* reductionism about tense
Thomas, Dylan 1
time; *see also* block universe theory of time; branching time theory; growing block theory; moving spotlight theory; passage of time; presentism
   instant of time, definitions of 143, 146, 151–2
   A-time 48–9
   B-time 48–9
   times as regions of spacetime 7–8
   times as abstract objects 39
   times as perspectives on reality 61, 62
time-like dimension 51–2
time-like line 133
Tooley, Michael 7n
Torre, Stephan 20n
truth-condition 14–15; *see also* MST-Supertense; MST-Supertime; MST-Time; reductionism about tense
twin paradox 136

unit of measurement, *see* quantity

Van Cleve, James 83n, 87n, 95–9, 129
van Inwagen, Peter 37n, 83, 114–15, 122, 126, 191n
visible property 194

welfare hedonism 234; *see also* tensed welfare hedonism
Weyl, Hermann 37
wholly present, *see* location, spatiotemporal
Williams, Donald C. 36, 42n, 51n
Williamson, Timothy 26n, 42n, 76n
Wüthrich, Christian 173n, 174n

Yablo, Stephen 219n
Yalcin, Seth 21n

Zimmerman, Dean 11n, 12n, 59n, 83, 145n, 149n, 150n, 179n, 235n

Printed and bound by CPI Group (UK) Ltd, Croydon, CR0 4YY